WHERE HAVE ALL THE SHEEP GONE?

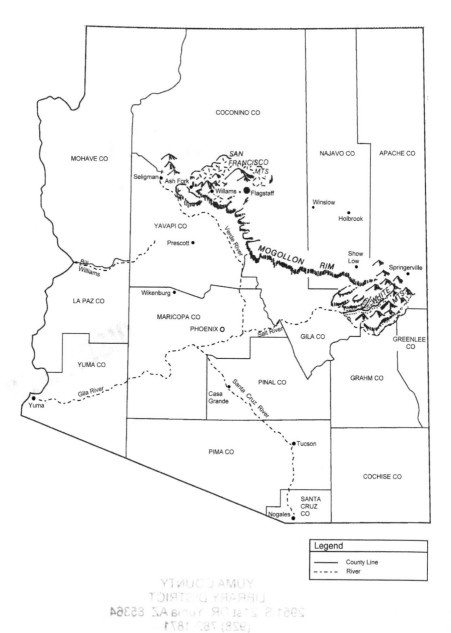

COCONINO CO

SAN FRANCISCO MTS

NAJAVO CO   APACHE CO

MOHAVE CO

Seligman
Ash Fork
Willams • Flagstaff
Winslow
Holbrook

YAVAPI CO

Prescott •
Verde River
MOGOLLON RIM
Show Low
Springerville

Bill Williams
WHITE MTS

LA PAZ CO

Wikenburg •

MARICOPA CO

PHOENIX O
Salt River
GILA CO
GREENLEE CO

YUMA CO

Gila River

PINAL CO
GRAHM CO

Casa Grande
Santa Cruz River

Yuma

PIMA CO

Tucson •

COCHISE CO

SANTA CRUZ CO
Nogales •

Legend	
——	County Line
- - - -	River

WHERE HAVE ALL THE SHEEP GONE?

Sheepherders and Ranchers in Arizona— A Disappearing Industry

by
Barbara G. Jaquay

Wheatmark Publishing

Where Have All the Sheep Gone?
Sheepherders and Ranchers in Arizona—A Disappearing Industry

Published by Wheatmark®
2030 East Speedway Blvd, Suite 106, Tucson, Arizona 85719 USA
www.wheatmark.com

Cover design and interior layout by Micah Edel, www.mistermicah.com

ISBN: 978-1-62787-458-8 (paperback)
ISBN: 978-1-62787-459-5 (ebook)
LCCN: 2016957984

2017rev01

ACKNOWLEDGMENT

———◆———

There are many individuals who have contributed to making this book possible. First and foremost, would be my friend, Beth Madison, who suggested the idea over a cup of coffee at our favorite coffee shop. Throughout this project, she has given suggestions, constructive criticism, and acted as a sounding board for organization of the book. Irene Aja opened her home and all her notes that she had compiled over the years on an industry very near and dear to her heart. Without Irene, this book would have been more difficult to write. She gave the author contact information for families whose parents were sheep herders and then many who became sheep ranchers and those who had been in the sheep business still living in the state. Through these contacts, I could interview many of the family members and write the history of the sheep industry. These families gave the author many of the pictures found within the covers of this book. The staff at Northern Arizona University Cline Library Special Collections were helpful and supportive of this project. They took the time to help me track books, documents and photographs within the Arizona Wool Growers Association files and other sources within their collection.

The staff in the Arizona Room, Phoenix Public Library, were also helpful in tracking documents and books.

Robbin Robinson contributed pictures of the Navajo-Churro breed and edited the chapter "Native American Sheepherders and Weavers in Arizona." Francisco J. Caraveo made the maps for the book. Individuals that contributed photographs include Carmen Auza, Melanie Aja Lanford, Cindy Shanks, David W. Schafer Photography, and Barbara Trainor Photography.

I want to thank Carmen Auza and Joe Manterola for allowing me and my husband to spend time with them and follow their sheep, documenting the history of their families, photographing activities undertaken by sheep ranchers each year, and that also would have been reminisce of past sheep herders and ranchers.

Finally, I wish to thank my husband for following me across the state to various libraries, down dirt and gravel roads, stopping constantly for me to take another picture, and who accompanied me on the interviews necessary to tell this story.

TABLE OF CONTENTS

FOREWORD

Where Have All the Sheep Gone? is an excellent example of research and writing about an important part of Arizona's history that has disappeared from the landscape. Growing up in Arizona, the movement of the sheep was a way of life. We looked forward to flocks moving from field to field in the wintertime and stopped to see and photograph the lambs in the spring. Trips to Flagstaff and other spots in northern Arizona we could see the sheepherders and dogs and marvel at the carvings the herders left on the Aspen trees.

The comprehensive story of the sheepherders and the ranchers to the state of Arizona is a story that has not been told before. I will recommend this book to all long-time Arizonians and those that have just moved here.

— **Carole De Cosmo**, Founder and Executive Director
Arizona Farming and Ranching Hall of Fame
PO Box 868, Glendale, Arizona 85311

Sheep on the White House lawn. Aug. 27, ca 1919.
Image. Retrieved from the Library of Congress,
https://www.loc.gov/item2004667793/ (Accessed October 22, 2016.)

INTRODUCTION

———◆———

There are many interesting tidbits about the sheep industry in the United States. For example, during the Wilson administration, a small flock of sheep from George Washington's Mount Vernon grazed on the lawn of the White House. The ninety-eight pounds of wool from those sheep sold in 1918 for a whopping $52,000 and the proceeds went to charitable groups. A little-known fact about the livestock industry in Arizona is that over a million sheep used to roam the grassy areas of mostly northern Arizona. Herding sheep was an important part of the economic activities that helped build Arizona during the early territorial days and as it gained statehood in 1912.

Today, just over 180,000 of the woolly creatures are in Arizona, both owned by non-Native Americans and Native Americans, mostly the Navajo. Most of the sheep are also found on small farms. The Navajo viewed sheep as a source of food and fiber for clothing. Sheep were seldom sold on the reservation. Sheep raised by non-Native Americans were always a commercial enterprise because the wool, a renewable resource, and the meat were sold for profit.

Many fortunes were made in the sheep industry, and during economic downturns or other disasters, some lost everything that they had worked so hard for. Many sheep ranchers hung on during the tough times, weathering the droughts and economic downturns. With the completion of the railroad across the country and feeder lines built to the central portion of the state, a new economic activity presented itself to the sheep rancher: a cheap means of shipping the lambs to eastern markets for meat. Wool was no longer the only way for the sheep rancher to make money.

The Native Americans were not all sheepherders as the Navajo and Hopi are today. Many herds were purloined from the early stocks brought into New Mexico and from Father Kino, who brought sheep into his Arizona missions. The Apaches were never interested in herding sheep; they took the sheep for immediate consummation. In Arizona, this led to a late start as new sheep stock had to be acquired with men bringing them in from California, Nevada, and New Mexico. By the 1890s, sheepherding was a major enterprise in Arizona. One feature of the industry was the practice of transhumance or the moving of sheep between winter and summer grazing lands, a common practice in the western states.

From the mid- to late 1900s, sheep were a common sight in parts of the Phoenix metropolitan area, grazing in the alfalfa fields during the winter months. Families would take their children to see the sheep in the alfalfa fields. Children were photographed gazing at the new-born lambs. Many of the old timers remember driving on the then-back roads of the Phoenix metropolitan area and coming upon hundreds of sheep that would swarm their cars as the herders moved them between different pastures. Others told the author about seeing thousands of sheep being moved along city roads as the sheep were moved to their summer grazing areas in the mountains of eastern and northern Arizona. Once no traveler thought much of the inconvenience of the sheep moving between fields in the metropolitan area or as herders and dogs began the long trailing process of moving the sheep to northern pastures, but those days are long gone. The last major sheep drive took place in the Phoenix metropolitan East Valley in 2011. No longer are there hundreds of sheep in the West Valley; today one can find sheep grazing year-round in only a few places.

Why has there been a decline in the sheep industry today? From 2012, Arizona agricultural statistics say 180,000 sheep were in the state, and they were raised on about 7,500 ranches with one sheep to over five thousand sheep. Only a handful of ranches raised sheep in the one thousand to over five thousand categories. The majority of the sheep are raised on approximately 5,700 ranches, and these have twenty-four or fewer sheep. The people who train sheepdogs own some of these small ranches; thus they need the sheep for training the various breeds of dogs that are then sold to sheep ranchers across the West. The wool production for the state in 2012 was 585,281 pounds, down from what was produced in 2007. The number of ranches has increased by about 2,500 from 2007 to 2012. The number of ranches or individuals raising sheep is on the rise today and shows two facts. A number of sheep are still in the state, and families find it profitable to raise large herds. While Arizona was never ranked high compared to other states in the total number of sheep or the amount of wool production, today it is still ranked as the eleventh leading state in the total number of sheep and lambs, and three of its counties rank within the top twenty of all counties within the United States.

The sheep industry needs to be captured before it is gone forever, as it's a rich history that has not been told comprehensively. This, then, is the untold story of the sheep industry in Arizona. It is the story of the many families working diligently through the late 1870s building the industry. Some of those families are still making their living raising sheep, trying to preserve a way of life that may be lost to those in Arizona in a short time. It doesn't matter what nationality the shepherd or rancher was—French-Basque, Spanish-Basque, non-Basque Spanish and French, Mexican, Peruvian, Dane, etc.—they all shared a unique way of life. Most of the research has shown that there was a pride in this way of life, especially in the Basque community.

It is hoped that the reader will gain a clear picture of conditions that existed for the early shepherds in Arizona and the changes that have taken place in the last thirty years. This is not a story that can be told in pictures even though it is said that a picture captures a thousand words. A way of life is disappearing, and once it is gone, a part of the unique history of Arizona will have gone with it.

Four terms are used in this book and are defined here to facilitate a better understanding. *Lamb* refers to sheep meat from sheep that are less than a year old, *mutton* is meat from sheep older than one year, *sheepmen* refers to men who owned the sheep and a *sheepherder* were the men hired to herd the sheep.

CHAPTER ONE

EARLY ARIZONA SHEEPHERDING

This is the story of the sheep ranchers and herders, soldiers, and missionaries who first forged a path into what is now the state of Arizona. The journey began in the 1500s and continues into the present day. Raising sheep in Arizona can best be looked at by dividing the history into three main periods. From the late 1500s to the 1850s, the historical record is sporadic about sheep, with most information obtained through military or missionary accounts written during the time. The second period is Territorial Arizona, 1860s to 1912, with the story told through accounts from newspapers of the time and the files of the Arizona Wool Growers Association (AWGA). The final period is from statehood to the present. It is the most detailed as many families who participated in the industry are still available to interview and tell their stories. While there will be some overlap between the second and final period, this is due mostly to families that started sheep outfits in the second period, continuing the tradition into statehood and beyond. Each of these periods will be dealt with in separate chapters.

Late 1500s to 1860s

Sheep were brought into the Americas first by Christopher Columbus on his second voyage and later into Arizona by the conquistadors. The earliest record for sheep that may have come into what would be called Arizona was with Coronado, when he crossed the southeastern corner of the state in his pursuit of the Seven Cities of Cibola in 1540. He had many animals with him, but what stands out is the five thousand sheep. It is also interesting that only twenty-eight animals actually made it to their final destination. Sheep supplied the men and priests with a food supply. The two priests who were to see to the conversion of the natives were left with the sheep, but, six months after Coronado left, the natives killed both priests, and the fate of the sheep has never been determined.[1]

Sheep were again brought into the United States when soldiers, Indian servants, and three Franciscans set out from Mexico in 1581. The natives killed all these padres too. It is interesting that when Espejo journeyed into New Mexico to rescue the padres, he was told that the native women were in the mountains with their flocks. However, it is just speculation that the animals were sheep. Towne[2] has stated that there could have been sheep as archaeological records indicate sheep were found in the old pueblos of the Rio Blanco in the Texas

---

1 Charles Wayland Towne and Edward Norris Wentworth, *Shepherd's Empire* (Norman: University of Oklahoma Press, 1945), 15.
2 Ibid., 29.

Panhandle. Specimens of wool and woolens were found at the pueblos. Coronado, traveling faster with a few soldiers, left the majority of his sheep to follow with some of the soldiers and servants, and these sheep could have been the ones that disappeared along the way and found their way to these pueblos.

With his colonization of the Rio Grande Valley, Juan de Onate brought a large flock of sheep into what is now New Mexico in 1598. Onate brought with him Churro[3] sheep, and these animals did well on the trail. It is from these sheep that the natives, such as the Navajo, obtained their flocks. There is no documentation to support that the Navajo were herding sheep by the late 1600s. Their sheep most likely came from pilfering raids in New Mexico. The Navajo have continued raising sheep up to the present time. Sheep were a very significant part of the everyday activities at missions in the Southwest and helped persuade the natives to convert. The only animal that provided both clothing and food was sheep. The natives were taught livestock husbandry and how to shear the sheep, card and spin the wool, and make cloth.

Through the records of the missions we find documentation of sheep once again, as all missions' economic well-being was based on sheep.[4] The Hopi pueblos of

---

3 Breeds of Livestock, Department of Animal Science, "Breeds of Livestock-Navajo-Churro Sheep" *Oklahoma State University*, Updated October 22, 1996, www.ansi.okstate.edu/breeds/sheep/navajochurro. The Churra sheep were brought into the New World and the word was later corrupted to Churro by "American frontiersmen," The Churra was an ancient Iberian breed known for its "remarkable hardiness, adaptability and fecundity."
4 Towne and Wentworth, 29.

Awatobi and Oraibi saw the first sheep brought by the Jesuit missionaries.[5] The first to be credited with breeding sheep in southern Arizona is Father Kino, who stocked his missions, visitas, and rancherias in the late 1600s.[6] On his trek into what is now southern Arizona in 1691, Father Kino brought in cattle and sheep for the beginnings of his missions of Guevavi[7] and Tumacacori, north of Guevavi. He had a thriving flock at Nuestra Señora de los Dolores, from which he could supply the many missions that he would begin in the twenty years he lived and worked in the Pimería Alta.[8] Sheep would provide the Native American with both food and clothing as they learned to breed the animal under the watchful, helping hands of Father Kino, and their herds increased, giving them ample animals for food and fiber.

Another missionary who worked with Father Kino, Father Saeta, also brought sheep into Tumacacori in 1695. The number of sheep brought into either of these early sites was not recorded, but Father Kino wrote in his diary that he brought in forty sheep to San Xavier del Bac in 1692. In 1700, when Father Kino was on one of his treks into Arizona, he wrote in his diary that there were forty sheep and goats at San Xavier del Bac and

---

5 Ibid., 39.

6 A mission had a resident priest. A *visita* was visited regularly by a priest and would have a small chapel, and a *rancheria* was visited infrequently by the priest. It had a place for Mass to be said, but its main function was the raising of livestock.

7 Guevavi is located north of present-day Nogales, along the upper Santa Cruz River. Today it is part of Tumacacori National Monument.

8 Pimeria Alta is an area of northern Mexico and southern Arizona that Father Kino was assigned for conversion of the native population.

eighty-four at Guevavi. When Father Saeta was murdered, the sheep and goats from his mission at Caborca were tracked to Tumacacori. Most of the early reports of sheep in Arizona were lumped together with goats, so there is no accurate figure for the number of sheep that could be found at some of the missions. There are a few incidents of only sheep being recorded, such as the two hundred sheep reported at Guevavi in 1701. It would be from these stocks that Kino would draw his supplies for further treks into the northern portion of Arizona. The main mission was always to be the primary stocker of cattle, sheep, and goats for the next place the padres would begin to work among the natives.[9]

Captain Manje, who was assigned to accompany Father Kino on many of his journeys, wrote of a trip where he, the padre, and natives were returning from the Colorado River via San Xavier del Bac. After leaving there, Father Kino became ill and sheep's broth revived him. The weather was likely responsible for his illness as "there rose such a furious hurricane of wind and rain."[10] Father Kino would seem to recover and the group moved on, but near Tumacacori, he once again fell sick. With the Santa Cruz River swollen from the rain, the natives brought over a sheep "to provide broth for Father Eusebio."[11] While the story does not add any historical significance to the sheep industry, it does show that the

9 Towne and Wentworth, 39–47.
10 Herbert Eugene Bolton, *Rim of Christendom: A Biographer of Eusebio Francisco Kino Pacific Coast Pioneer* (Tucson, AZ: University of Arizona Press, 1984), 423.
11 Ibid.

native population thought that the sheep had medicinal powers.

Sheep are mentioned sporadically in the literature for the next 150 years. Kessell[12] wrote that a census was required at the missions of the Pimería Alta. He states that in 1768, three hundred fifty sheep were at Guevavi. Padre Escalante reported that the Hopis had thirty thousand sheep in 1775.[13] Padre Font reported seeing sheep with the Pimas along the Gila River in 1776.[14] Padre Garces also reported seeing sheep among the Hopi in 1776, but by 1780, after a severe drought, only three hundred sheep were spotted.[15] The flocks brought initially into Arizona would survive and thrive, hindered only by the Apache, who would attack and take the sheep for food, or the Navajo, who took them to raise for both food and fiber. This leads us into the beginnings of the sheep industry in Arizona.

Sheep Drives to California

During the gold rush days in California, sheep numbering in the hundreds of thousands were driven from New Mexico to California. The men followed two routes. The northern route followed the Santa Fe Railroad right-of-way and was probably the easiest route to follow. Kit

12 John Kessell, *Friars, Soldiers, and Reformers: Hispanic Arizona and the Sonora Mission Frontier, 1767–1856* (Tucson, AZ: University of Arizona Press, 2016), 52.

13 Herbert E. Bolton, *The Padre on Horseback A Sketch of Eusebio Francisco Kino, S. J. Apostle to the Pima.* (Chicago, IL: Loyola Press, 1982), 65.

14 George Peter Hammond, *Don Juan de Onate and the Founding of New Mexico* (Santa Fe, NM: El Palacio Press, 1927), 180.

15 Elliott Coues, *On the Trail of a Spanish Pioneer: The Diary and Itinerary of Francisco Garces,* (New York, NY: F.P. Harper, 1900) II, 361.

Carson reported on his last trip to California out of Santa Fe that nearly a hundred thousand sheep were on the road.[16] The southern route was the toughest of the two routes. The trail used by the sheepmen began in Las Cruces left the Rio Grande, leaving the flock and man without water for sixty miles. North of present-day Deming, water was found at the Rio Mimbres. Then heading to the "Animas Valley, through Guadalupe Canyon, into Arizona and San Bernardino Springs, south into Sonora, Mexico, along the Lamorita to a point south of Bisbee, Arizona, west to the San Pedro River, over the Divide, then north to the Terranate, the Santa Cruz, and the Gila River to the Colorado and the California line at Yuma."[17] The United States Department of Agriculture estimated the number of sheep driven across the state to be well over five hundred and fifty thousand sheep.[18] Half of these sheep traveled the southern route.

John Abel and his partner, Benjamin Riddell, led one of these drives using the southern route in 1855. With ten thousand sheep, the men started from Chihuahua, Mexico, and headed to California. They were accompanied by Jose Castañeda, the master of ceremonies of the commissary department. Driving the sheep across what would become the Arizona and Mexico border, they camped for

---

16 W. W. H. Davis, *El Gringo, or New Mexico and her people.* (New York, NY: Harper & Bros, 1857), 125.

17 Towne and Wentworth, 93.

18 U.S. Department of Agriculture, Bureau of Animal Industry, *Special Report on the History and Present Condition of the Sheep Industry of the United States,* prepared under direction of Dr. D. E. Salmon, Chief of the Bureau of Animal Industry, (Washington, D.C.: Government Printing Office, 1892) Part II, 919.

about six weeks southeast of Bisbee on land owned by John Slaughter. There, Miguel Mangas Coloradas, along with three hundred men and warriors, women, and children, entered the camp, showing friendship. Abel and his men gave them clothes and ten sheep. The Native Americans left the next day after a feast. It wasn't a week when twenty-five bucks and two chiefs reappeared, indicating that they were hungry, but they showed their displeasure when they were given only two sheep. The Native Americans took the sheep and poisoned them, leaving them in a nearby canyon. When the poisoned sheep were found, Abel and his men knew that trouble was brewing. This trouble began about two weeks later when five braves sneaked into the camp, stole five horses, and stampeded the flock. The sheep were scattered over a large area, but were soon rounded up, and they considered their losses of five horses better than what it could have been. After these two encounters, they broke camp and headed to the Santa Cruz River Valley to just south of Tucson and then continued up to the confluence of the Gila River. They followed it until they reached Fort Yuma. There, the flock was ferried across the river west of Yuma at Santa Cruz. Tragedy struck again as about a thousand sheep died when they ate a poisonous herb growing on the banks of the stream. Upon reaching Los Angeles, they sold most of the herd for $8 to $10 apiece as well as their wagons, horses, and mules—anything not needed for the return trip. The rest of the sheep were sold in San Francisco. The men made a good profit in both locations since the sheep had cost them about fifty cents each in Chihuahua. It was common to find sheep selling for $16 a head at the beginning of the 1850s, but by the end of the decade, the price was down to $3.37 per head, still a profit if one could purchase sheep at fifty cents per head.[19]

Others who herded sheep across southern Arizona and attempted to cross at Yuma were successful in crossing here, but some were not so lucky. Two unlucky participants in the drives were John Gallantin and Josiah White. Gallantin had been a bounty hunter for Apache scalps for the Mexican government. Not all the scalps he turned in were Apache—some were friendly and innocent Papagos and Yaquis and some were Mexicans. On this particular trip, he had both purchased and stolen his flock of two thousand five hundred sheep from New Mexico. At Yuma, the Yuma Indians murdered him, along with twenty-five other Americans herding their sheep to California.[20] White had four thousand two hundred seventeen sheep that were taken by the Native Americans when he and his crew abandoned them to reach water to save themselves as the sheep had refused to move. Mr. Croon was one who successfully crossed the Yuma with fourteen thousand sheep. He had forty-five Americans and fifteen Mexicans with his livestock to ensure the safety of the outfit.

---

19 Ibid.

20 John C. Cremony, *Life among the Apaches* (Glorieta: NM: Rio Grande Press, 1969), 116.

Another group that made the trek safely was with Felix Aubrey. Aubrey was considered the first explorer of a wagon route over the thirty-fifth parallel when he showed that a wagon could be driven between San Jose, California, and Santa Fe, New Mexico. The route would become the Atchison, Topeka, and San Francisco Railway. On at least one of two trips into California in late 1852, Aubrey and sixty men started with three thousand five hundred head of sheep, but the number increased with lambs born in route. For his second trip in 1853, he gathered fourteen thousand sheep, but before he left, he was joined by others until a herd of fifty thousand sheep was amassed. The trip took three months, with no incidents reported of trouble along the route. It was reported that Aubrey made a fortune selling the sheep in California.[21] Other reports about the movement of sheep could have been cited also. These accounts show the enormous number of sheep that were moved across the state in the 1850s.

Military Excursions across the State

Military surveys in northern Arizona brought sheep for feeding the men. Lieutenant Whipple in 1854 was the first, followed three years later by Lieutenant Beale, who was famous for using camels to move across the desert. Whipple and Beale had learned from Captain Sitgreaves,

---

21 Some of this information comes from several sources, such as Haskett, by an unidentified author found in the Arizona Wool Growers collection in the Cline Library, Northern Arizona University, and Thomas Farish's eight-volume *History of Arizona*, 1915 to 1918, that contains many personal manuscripts of people who lived during the time and early documents.

who, in crossing northern Arizona in 1851, had to slaughter his pack mules to keep his crew alive.[22] While these sheep were not heading to California, a few probably were not used for food, and some may have been pilfered by the northern Native Americans, such as the Navajo. As a side note, Lieutenant Beale would raise sheep on three ranches in California upon his discharge from the military.

When the Mormon Battalion began its march across New Mexico to California, sheep were not initially included. It would be two weeks into the march before three hundred sheep were purchased near Socorro, New Mexico. The sheep were considered poor quality; thus another eighty were purchased the next day. Unfortunately, an army the size of four hundred soldiers would eat a cow and twelve lambs per day. There was no way that the sheep would be able to supply the men for the entirety of the trip! The battalion ate abandoned cows and bulls as they marched across Arizona, and it was reported that they still had 130 sheep when they reached the Colorado.[23]

While sheep crisscrossed the Arizona territory, few sheepmen or even cattlemen would drive their livestock here until the Apache problem was solved. Prior to the United States acquiring Arizona through the Mexican War of 1848 and the Gadsden Purchase of 1854, the Apaches attacked Mexican settlers and stole their livestock, an easy

---

22 Bert Haskett, "History of the Sheep Industry in Arizona," *Arizona Historical Review* 7 (1936), 11.

23 Towne and Wentworth, 108–109.

food source. As was stated earlier, the natives left some sheep at the ranches to become stock for future raids. With the arrival of the American military, the Apaches were mostly kept at bay, but when the soldiers were pulled to fight in the Civil War, the Apaches had little restraint and attacked the settlers, stealing their livestock and killing many of the settlers. According to Cremony,[24] they had a unique way of moving the sheep. The Apaches first would move the flock far enough after the raid to elude capture by anyone attempting to pursue them. The sheep were then organized into a parallelogram, with the strongest sheep lashed together in pairs to form the sides or a fence-like structure made out of the animals. The weaker sheep would be found inside this enclosure. This forced them to keep up with the rest of the animals. Warriors would be placed around the animal enclosure and would keep the animals moving. This method allowed the natives to move animals long distances. At night, this fence would keep any animals from straying.

The Apaches, with no one to stop them, were able to raid further westward. However, in one raid, the Apaches did not count on being chased. They stole two thousand sheep not far from Fort Whipple, near present-day Prescott. This raid did not go in favor of the Apaches, as peace officers were able to overtake the raiding party and recover the sheep of L. A. Stevens and Levi Bashford, who were headquartered near Granite Creek. It

---

24 Cremony, 282.

can be surmised that the Apaches had no time to form their traditional method of moving the sheep.

There was some settlement in the northern portion of the state during this period. Jacob Hamblen drove sheep into the Little Colorado Valley in the 1850s, after crossing the Colorado River at Lee's Ferry. On one trip in 1858, he saw a Hopi sheep corral. On subsequent trips, he traded with them and gave them sheep shears and wool cards.[25] Hamblen may have noticed that the Navajo and probably the Hopi sheared their sheep with a knife or piece of sharpened iron and that the sheep shears would be better for shearing.[26] Other Mormons would follow Hamblen and settle along the Little Colorado River. Their flocks increased to the point that it was warranted to build two wool mills at Tuba City and Moencopi. Both mills were short-lived for neither one could compete with California woolens that were "readily available in the market."[27]

25 Juanita Brooks, "Jacob Hamblen," *Arizona Highways* 19:4, (1943), 32–33.

26 U.S. Department of Agriculture, Bureau of Animal Industry Part II, 943.

27 Edward Norris Wentworth, *America's Sheep Trails*. (Ames, IA: Iowa State College Press, 1948), 242–243.

CHAPTER TWO

THE TERRITORIAL YEARS: 1860s TO 1912

Sheep herded across Arizona to California in the 1850s would now be herded back across to New Mexico in the 1870s. The sheep fetched a substantial profit as they were purchased for $2 a head and sold for $3.50.[28] Many men purchased these sheep in Merced County, California, and followed the Beale Wagon trail across northern Arizona, a route that would become synonymous with Route 66. It took up to seven and a half months to herd these sheep from California to New Mexico, but not all sheep were herded across Arizona.

Two reasons explain why sheep would reverse directions. First, a series of droughts in California forced many cattlemen to drive their large herds to Arizona to graze on the green pastures not affected by the droughts. Sheep would have a profound effect on the Arizona ranges. They carried alfilaria in their wool and on their feet. Alfilaria became the most valued feed on the Arizona ranges. Cattlemen, who in future years would oppose the sheep, would be thankful for the sheep bringing this valuable feed source into the state.

---

28 U.S. Department of Agriculture, Bureau of Animal Industry, 1892, Part II, 929.

The second reason sheep headed to New Mexico was to improve their breeding stock. The sheep in California at this time were better stock, and the breeds in New Mexico needed improvement.

Droughts in California

While marauding Apaches stole livestock and killed settlers in Arizona and New Mexico, California was free from warring Indians. Livestock at the California missions flourished. The Native Americans were taught animal husbandry and the art of weaving clothes from the wool, just as Father Kino had done in the missions in Arizona. Sheep provided a food source in the form of milk and meat. Raising livestock was an activity outside of the missions too, but, climatic conditions would change the California livestock industry. A severe drought lasting twenty-two months occurred from 1826 to 1830. Large numbers of animals were slaughtered as the animals starved from the lack of grass and water. Some missions sacrificed their horses and mules to keep what little grazing area there was for the cattle and sheep. Another drought occurred in 1840. Again, rain did not fall for fourteen consecutive months. Sheep starved or were killed by hungry predators. Sheep at the missions and elsewhere in California dwindled in number.[29]

The California drought in the early 1860s was particularly severe and forced the men to look for green

---

29 Towne and Wentworth, 49, 51. Haskett, Bert, 22.

pastures. Getting the sheep to the Arizona territory also had its challenges as one early writer reported in 1903 that the wrong type of forage and the scarcity of water as they crossed the desert added to the already difficult circumstances that the sheepmen had experienced in California. The animals were weakened from the lack of good grazing lands, and then they had to contend with the conditions along the trail. Once the men and sheep reached the Colorado River, the sheep had to be ferried across the river. This posed another problem: prices for hauling the sheep was set by the ferrier and could be steep. One incident stated in *The Arizona Weekly Miner*, June 25, 1875, the sheepmen were not always ripped off by the ferrymen. Peter Filance wrote:

"...we drove from California 4,000 head of sheep over the toll-road known as the Hardyville Road, and the Toll-gate man charged us $800. This amount we thought was more than sheep man could afford in their efforts to settle this country. We called on Mr. Hardy and entered our complaint after his arrival from California and he voluntarily paid us back $400. We feel it our duty to say in behalf of Mr. Hardy, that he acted very gentlemanly with us, and we feel as though we could say to those traveling from California that they will find in Mr. Hardy a gentleman and a man that will deal fairly and honorably with them."

The few early histories written about the sheep industry do not always agree as to some of the dates and details of those first sheepmen into the territory. Some early records come from newspaper accounts. In the early 1870s, during the California drought, a large herd of sheep was driven into the Tubac area in southern Arizona. While giving no source for his information, Haskett stated that "two Basques from California grazed their herds in the vicinity of Oracle, Pinal County. Later accounts of them, however, are lacking."[30]

Southern Arizona

In the 1870s, a number of haciendas, including the San Bernardino and Babocomari, were established in southern Arizona, with reports of large flocks of sheep as well as cattle. Antonio Aros brought sheep and horses across from California, crossing the Colorado River into Mexico, where he then herded them north through the Altar Valley. His livestock grazed on the eastern slopes of the Baboquivari Peak area.[31] Others were established along the San Pedro, Santa Cruz, and Sonoita Rivers when Spanish settlers arrived from Mexico. Many of the sheepmen are unknown in this part of the state, and the number of sheep comes from a promotional book sponsored by the Arizona Territorial Legislature in 1881. Pima County was said to have fifty thousand sheep and Pinal County, two thousand. While this report knew of sheep in Cochise, it recently split from Pima County, and the county government did not have the exact numbers of livestock of any breed.[32]

---

30 Haskett, 20.
31 Wentworth, 261.
32 Patrick Hamilton, *The Resources of Arizona*. (Territory: Prescott, 1881), 90.

While raids by the Apaches were more common in the southern portion of the territory, sheep were still grazed there. It was reported in the *Arizona Citizen*, May 17, 1873, that Charles Marsh, of Marsh and Driscoll, "grossed nine hundred dollars from four hundred ewes in two years." (That is over $19,000 in 2015 dollars, adjusted for inflation). A flock of ten thousand sheep were brought into southern Arizona from New Mexico in 1874 and put on the range west of present-day Arivaca. Tully ran the sheep and his firm, Tully, Ochoa, and Company, also ran an extensive freighting business out of Tucson.[33] The outfit was never harassed by the natives, and at the time, it was unclear why they had no Indian troubles even though the freight was carried through territory that Geronimo consistently attacked. In an article in the *Arizona Citizen*, February 17, 1940, the mystery was solved when it was learned that Ochoa had saved the life of a chieftain under Geronimo. The lack of reliable water and shallow wells being no deeper than six feet doomed the Arivaca area in the summer. In other words, it would dry up quickly in hot weather. The *Arizona Citizen*, April 4, 1874, disclosed that the outfit brought in another flock of sheep; this time four thousand, five hundred from New Mexico, and placed them on a range east of Tucson. Pedro Aguirre ran both sheep and cattle in the Arivaca area. Aguirre brought in an unknown number of cattle and sheep to take advantage of the lush grasses on the hillsides and established the Buenos Aires ranch.

33 Wentworth, 248.

Several editions of the 1874 *Arizona Citizen*, April 4, May 16, May 23 and December 19, announced more sheep coming into the Gila River area from California in 1874. Flocks of sheep were no longer pilfered by the Pima Indians (the Tohono O'odham today) as they were placed on a reservation. The governor of the territory, Governor Safford, and the Carr brothers brought in an unreported number of sheep. Later in May, Charles Horn brought in one thousand five hundred sheep for Governor Safford. Safford brought in highly bred animals in December of that year—two Spanish Merino bucks and thirty head of Spanish bucks and ewes. Each group was purchased from different breeders in California. Others followed that same year, bringing sheep in from both California and New Mexico. This included G. W. Hance, who brought four hundred sheep from New Mexico into Pima County, and George H. Stephens and Lord and Williams, who brought them from California. Two thousand sheep came with Williamson and White, and one thousand three hundred were brought by Maloney, Craft, Garrett, Steele, and Anderson. Mr. H. Maloney wrote a letter in Los Angeles to the *Arizona Citizen* and stated that thousands of sheep would be soon driven into Pima County. Mr. Maloney named the abovementioned individuals, who would be bringing these sheep into the county as of course he was part of the latter group. A few months later, the *Arizona Citizen*, April 3, 1875 reported that the men arrived with the one thousand three hundred head of sheep near Tubac.

In travels through the Arizona territory in 1874 to 1876, Hodge stated:

"The section of the country watered by the Chiquito Colorado is especially favorable to sheep raising, as is also the region of country around Prescott and thence to the north in the region of... the San Francisco and Bill Williams Mountains. The same can also be said of the country south of Tucson embracing... the Santa Rita, Patagonia, Huachuca, Whetstone, Dragoon, and Chiricahua Mountains and in the contiguous valleys."[34]

Hodge further reported that in 1875 sixty tons of wool was shipped out of Yuma. No other information is given on this shipment, so we do not know where the wool originated. It could have come from the northern portion of the state, but that is unlikely considering the distance. Before the completion of the Atchison, Topeka, and Santa Fe Railroad across northern Arizona, the northern growers used the Colorado River to transport their wool to eastern markets. Hodge's statements showed that at least by 1876, the sheep industry was well underway in the southern portion of the state and fear of raiding natives had lessened due to the strong presence of the military that had returned after the Civil War.

In 1881, Hamilton,[35] writing for the Arizona Territorial Legislature, stated conditions were favorable in the

---

34 Hiram Hodge, *Arizona As It Is*. Compiled from notes of travel during the years 1874, 1875 and 1876, (New York, NY: Hurd and Houghton; Boston, MA: H.O. Houghton and Company, 1877), 56.
35 Hamilton, 281.

Arizona territory for both cattle and sheep because the gamma grasses covered mountains, valleys, and mesas and the climate was not as fierce as that found in other western areas of the country. The good weather meant that in the winter months no protection was needed for the livestock. The Indian troubles of raiding livestock had ended with the creation of Indian reservations. The completion of the railroad made it easier to ship meat and wool. The grasses gave mutton a "fine flavor and tenderness."[36] The report went on to say that by ending of the high rates of railroad freight, "the sheep industry of Arizona is one of the most lucrative branches of business in the territory." The alfileria or wild clover introduced on the sheep's wool when brought into Arizona from California was doing well, and both cattle and sheep fed on it. Yavapai County had the richest grazing area in all the territory. The climate that brought heavy snows in the winter and rains in summer allowed a nutritious growth of grasses on which the livestock could feed. The southern portion of the territory, Pima and Cochise Counties, also fared well with the grasses, making the area along the Santa Cruz River, the valleys of San Simon, Sulphur Springs, and San Pedro, ideal rangeland. The same could be said of Graham, Pinal, Maricopa, and Gila Counties, which had thousands of acres not being used for grazing at the time of the 1881 report. It was stated:

---

36 Hamilton, 89.

Speaking in general terms, it can be truly said that there is no better grazing region west of the Rocky mountains than Arizona, and while the want of water prevents many portions of the country from being occupied, there is yet room for thousands of cattle and sheep where water is abundant, where animals keep fat winter and summer, where the climate is all that could be desired, where disease is unknown, and where an energetic man with a small capital, who understands the business, can make himself independent in a few years.[37]

The Hamilton's section on "Grazing" ended by stating that within the counties of Apache, Pima, Yavapai, Maricopa, Graham, and Pinal, there were 408,316 sheep, with a decrease in numbers from the former to the latter.[38]

A few woolgrowers in the Tucson area were D. A. Stanford, McGarey Brothers, J. J. Blake, Walter L. Vail, Sanford and Hilton and Tully, Ochoa, and DeLong.[39] D. A. Stanford foreclosed on a ranch with thirteen thousand head of sheep that had been owned by Tully, Ochoa, and DeLong. He sold off his cattle and began to run a profitable sheep business. While Stanford, without his partner, is the only one with even a brief mention, this at least shows that some sheep were in southern Arizona. Pima County in 1895 was reported to have 2,134 sheep at a value of $3,203.[40] If Stanford had thirteen thousand sheep in 1881, there is a question as to what happened to all those sheep in fourteen years. Many sheep are not accounted

37 Ibid., 90.
38 Ibid., 90.
39 Haskett, 23.
40 Edward H. Peplow Jr., *History of Arizona II*, (New York, NY: Lewis Historical Publishing Company, Inc., 1958), 379.

for. No satisfactory answer has been found for the difference in numbers.

The Southern Pacific Railroad was used in 1877 to bring sheep to Fort Yuma from unknown parts of California. Once the sheep owned by General Banning were unloaded from the train and rested a few days, they were trailed into the Santa Cruz Valley. Two other shipments of sheep via the same route would occur, one with 1,250 sheep and the other with 2,000 were reported in the *Arizona Weekly Star*, July 12, 1877.

It is also interesting to note that two cattlemen, W. H. Bayless of Kansas and J. W. Berkalew of New York, purchased Mr. Kay's three thousand head of sheep and several ranches in the San Catalinas for $10,000. The *Coconino Weekly Sun*, February 25, 1892 was told the men made the purchase to protect their stock ranges, which Kay's sheep were trespassing on.

While the exact number of sheep is unknown for the southern portion of the state, it was reported that 384,000 pounds of wool was shipped out of the territory on the Southern Pacific Railroad in 1892.[41] That same year, twenty thousand sheep were trailed from Texas and grazed in Sulphur Springs Valley in the foothills of the Chiricahua Mountains.[42] Pima County was said to be ideal for large flocks of sheep as "the great mountain ranges produce large quantities

---

41 U.S. Department of Agriculture, Bureau of Animal Industry, 1892, Part II, 943.
42 Ibid., 943.

of the nutritious grasses."[43] The report continued to say that the larger livestock would not be able to access the grass as easy as sheep could. Furthermore, it said that it was definitely an easier opportunity for a person with a small amount of capital to invest in sheep than in cattle, which required a large outlay of cash. The return on sheep would be realized sooner since there would be a wool crop yearly. As of the writing of that report, the "higher price of wool incident to the passage of the late tariff legislation by congress and the greater demand for mutton as an article of food, owing to the increased price of beef cattle"[44] was a benefit to sheepmen.

Eastern Arizona

Tonto Basin was the most important sheep grazing area in eastern Arizona with nutritious grasses[45]. The detail about the grazing conditions between about the 1870s and 1926, the date of the report is most fortuitous. This author stated that the lush grass area of the Tonto and on the west slope of the Mazatzals were used by the sheepmen to winter graze and to lamb in the spring. He added that unfortunately the area was fully stocked by the cattlemen and hard feelings soon developed.

The cattlemen surprised a herd that belonged to George Scott, whose herders and tenders had brought four bands

---

43   *A Historical and Biographical Record of the Territory of Arizona* (Chicago: McFarland and Pool, 1896), 1012.

44   Ibid., 1012.

45   Fred Croxen, "History of Grazing on Tonto." Presented at the Tonto Grazing Conference in Phoenix, Arizona, November 4–5, 1926.

of sheep into the area. The men, even though heavily armed, were all disarmed, and the sheep were shoved into Fossil Creek. The sheepmen were told not to return. George Wilbur was another sheepman who was run off the range. Unfortunately, there were no dates given for any of these encounters, but it had to be prior to the Pleasant Valley War.[46]

Juan Candelaria is credited with the first sheep being reintroduced into eastern Arizona after the Apaches had decimated the population.[47] He drove seven hundred sheep to a ranch site he had selected a few miles south of present-day Concho in 1866. He was followed in the next few years by his three brothers, Rosalio, Ambrosio, and Averisto, who also brought sheep from New Mexico and settled near Concho. This was five years before the Navajo were resettled back in northern Arizona after Kit Carson forcibly removed them in 1863. The Candelaria sheep were considered the finest sheep as they were directly descended from the original seventy-one sheep that their grandmother had purchased in Mexico and drove into New Mexico during the early 1800s. Of the original seventy-one sheep, three ewes and a black ram were all purebred merinos. In the two years, from 1799 to 1801, it took to drive the sheep from Vera Cruz to Cubero, New Mexico, the sheep had increased to one hundred and forty-one. That bloodline was never broken. Situating his ranch between the future Apache and Navajo reservations, Candelaria stated that he was never

---

46 Ibid., 1012
47 Haskett, 19.

once bothered by Native Americans from either reservation as "he knew Indians and was their friend always."[48] Candelaria was reported to be the largest sheep owner and taxpayer in Apache County up until the time of his death in 1930.[49]

Migratory sheep outfits from New Mexico grazed in the Little Colorado Valley between 1868 and 1870 and in the Springerville area during 1870. These two areas were rich in feed for the sheep, and water was not an issue. The Apaches attempted to dislodge these migratory outfits but were unsuccessful.

Again citing the promotional book sponsored by the Arizona Territorial Legislature of 1881, Hamilton lists three hundred thousand sheep in Apache County and thirteen thousand in Graham.[50] The majority of the sheep for Apache County had to be those owned by tribal members of the Navajo Reservation.

The last known man to bring sheep into this area was Isador Solomon in 1868. Solomon settled along the Gila Valley in Graham County.[51]

Prescott and the Surrounding Area

One of the first to herd sheep here was James Baker, who settled in Chino Valley near Fort Whipple. Baker

---

48 Ibid., 19.
49 Ibid., 19.
50 Hamilton, 90.
51 Haskett, 20.

drove an unknown number of sheep across the Mohave Desert, but lost a good portion of the sheep to native raids. By 1870, he had close to a thousand sheep grazing in the area. The local *Weekly Arizona Miner*, Prescott, March 4, 1871, referred to sheep happenings in the area. During the first part of March 1871, Campbell and Buffum sent an unreported amount of wool on consignment to San Francisco, for which they received twenty cents a pound. This may be the earliest report of an amount paid for wool produced in the Arizona territory. The September 2, 1871 *Weekly Arizona Miner*, reported Campbell and Baker brought in a flock of eight hundred sheep, and they had not been molested by the Native Americans. Again in 1874, the pair shipped four thousand pounds of wool to San Francisco and were paid thirty cents a pound. It should be noted here that much is unknown about Campbell and Buffum. We do know that this was John G. Campbell and that he partnered with both Baker and Buffum. There would be other Campbells, such as Hugh, Colin, and William who would come later into the sheep business, and they herded their sheep in Coconino County around Flagstaff.

The American Ranch owned by J. H. Lee was one of the earliest ranches to raise sheep and the best known during the 1870s. Lee arrived with others in 1864, making him the third party to settle near Prescott. The ranch became the center of operations for the area because the military could get their supplies, and it was a stage and mail stop. Lee and General Crook, commander at nearby Fort Whipple, came from the same township in the east, giving Lee a "lead-pipe cinch."[52] Much of the information about Lee and his sheep outfit comes from the local newspapers during 1875. Lee first added an unknown number of sheep to his ranch in 1871 and was a big promoter of the industry. He purchased a dozen well-bred rams to improve his flock in 1873. These rams, along with a thousand other sheep from New Mexico, were trailed into Arizona. Wentworth states that "Lee gained great prestige in the territory by slaughtering 'a Spanish wether'[53] in 1875 that dressed a fifty-seven-pound carcass."[54]

On January 29, 1875, Lee wrote a letter that was printed in the *Arizona Weekly Miner*, Prescott paper:

> "I have been engaged in wool growing for four years and am $1,600 out of pocket. I believe this is the experience of all engaged in sheep raising on this side of the mountains. That side (east) is the only suitable place I have seen in Arizona. All are getting over there—Foster, Frank Hart, and Smith Bros. are there; Joe Marlow is on the way, also Charlie Stevens with his band from Hackberry. I think the day is not far distinct when the sheep owner may become a respected man, not held out at arm's length, surveyed, and smelled of when he asks a favor."

Lee moved his sheep to the Little Colorado area in December 1877, where he planned to shear them in the spring with many other outfits. Shearing was about half the price in Coconino County versus the going rate in Yavapai. Wool prices were the same, whether they shipped to the

---

52 Orick Jackson, *The White Conquest of Arizona*. (Los Angeles: West Coast Magazine, Grafton Co., 1908), 45.
53 A castrated male
54 Wentworth, 247.

Missouri River or to San Francisco, even though San Francisco was closer.

Lee also had another problem: the natives. The natives ran off his only guard and burned the ranch, including cash that Lee had on hand. He obtained more collateral and rebuilt, restocking his livestock and growing crops. The natives again attacked, but the new lessee, while escaping with his life, left a bag of bad flour in plain sight; it was laced with a heavy dose of strychnine. Soldiers coming upon the station found the dead and very ill natives who had consumed the flour.[55]

Manuel Yrissari brought sheep into the Prescott area in 1872 when he trailed two thousand five hundred sheep across the Navajo Reservation. While not harassed by the natives, he did complain of the lack of water between Albuquerque and Fort Wingate (just east of Gallup, New Mexico) and his gratitude for the rainfall that supplied his needs.

In 1873, Lee and Yrissari were joined by W. A. Deering, who trailed three thousand sheep from the Mint Valley in California. In November of that year, Joseph Curtis brought in a thousand sheep from the West Coast. He had made his headquarters on Granite Creek, the same vicinity as Stevens and Bassford. In the *Arizona Citizen*, December 13, 1873, an unidentified sheepman was reported to have brought two thousand sheep from a California-bred flock to Kirkland Valley south of Prescott.

Cordes Station, located southeast of Prescott, was established in 1883 as an important stopping place for those trailing

---

55 Wentworth, 246–247.

their sheep along the Black Canyon Driveway. (The term driveway is used here to describe a narrow route by which sheep were moved from one location to another. It was restricted by the U.S. Forest Service by width). This important station will be described more in the chapter on Trailing.

A shift away from Prescott and into the Bill Williams Mountain begins as the sheepmen need better rangelands and a more reliable water source.

Northern Arizona and the Beginning of Flagstaff as Sheep Headquarters

John Clark was the first sheep rancher in the northwestern part of the territory.[56] Born in Maine in 1839, he spent some time in Massachusetts with a sister and then moved west, arriving in California after crossing the Isthmus of Panama. Being an able young man of twenty, he went to work at a variety of jobs, including a dairy farm, for the next few years, and his first introduction to sheep occurred. He was in charge of five hundred sheep but left and took up employment with an outfit in Merced that had between sixteen thousand and twenty thousand sheep. He bought his first sheep a few years later, and after believing that the Southern Pacific Railroad had spoiled the open range and after experiencing the drought in California, he left Kern County, California, and started for Arizona with five thousand sheep, bringing them in by crossing the Colorado River via Hardy's Ferry in December 1875. Weather, in the form of a

---

56 Peplow, 148.

snowstorm or a severe sandstorm, depending on the source one wants to rely on for information, hampered his movement of the sheep, and he lost over three thousand of them. He wintered that year near Big Sandy with the remaining sheep. He then moved to what is now Coconino County[57] to the area of Bill Williams Mountain. He made his last move in 1883, settling in Flagstaff. Some believe that he may have been Flagstaff's first settler.

Until 1887, he grazed his sheep north of Flagstaff in what is known today as Clark Valley, a narrow canyon situated between Elden and Mormon Mountains. Clark ran cattle for a time, and later, with his brother-in-law, George F. Campbell, ran sheep. Clark was a successful sheepman, and in 1883, he sold five thousand head of sheep for one dollar each, a high price for the time. In the Flagstaff area, he purchased 320 acres from his brother Asa, who had arrived in 1883. Asa had bought the property from Thomas F. McMillan and a home was built. John may have been able to buy the ranch, later known as Clark Ranch, from the large amount of money he got from selling the sheep. He grazed more than ten thousand sheep on the ranch. When Asa Clark bought the land from McMillan, they did not know that was the section designated for a school. Asa battled for four years to get title to the land before he sold the ranch to his brother. It would take many more years before John would be able to acquire clear title to the

---

57 Coconino County was not established until February 19, 1891. Flagstaff and Williams were both part of Yavapai County, one of four counties established when Arizona was given its own territorial status separate from New Mexico.

ranch. Clark was involved in many businesses and politics after he sold his sheep.[58]

Clark was soon joined by Ashurst and Blake. William Ashurst, born in Missouri in 1844, went to California with his parents in 1856. He worked in the mines until he married in 1871, and then the young couple would drive sheep across to Arizona out of Utah, wintering in the area of the Big Sandy River.[59] He pastured his sheep in what is today called Ashurst Run, but back then it was called Anderson Mesa. The couple had ten children and for many years lived at Anderson Mesa. He sold his sheep in 1882 and went into the cattle business. Ashurst Lake is also named after the family. Ashurst was the father of Arizona's Senator Henry F. Ashurst. John Blake, having success in California herding sheep in 1874, moved his operation to Arizona sometime thereafter but sold out in 1881.[60]

Early newspaper accounts told of a sheep ranch located at Johnson's Canyon, approximately ten miles west of Williams by the late 1870s. The *Weekly Arizona Miner,* May 12, 1876, mentioned Mr. Coleman, who brought four thousand head from Visalia, California, into the Williams area. The same newspaper reported on May 25, 1877, John G. Campbell may have been here grazing

---

58 Platt Cline, *They Came to the Mountain* (Flagstaff: Northern Arizona University, 1976), 91–92.

59 Oral History Interview with Senator Henry Fountain Ashurst [includes transcript] May 19, 1959 and October 9, 1961, University of Arizona Cline Library. The Ashurst Family manuscripts state that Clark and Ashurst crossed the Colorado River together.

60 *Portrait and Biographical Record of Arizona.* (Chicago: Chapman Publishing Co., 1901), 631. Blake was active in many things, and he was a probate judge for four years.

sheep near Bill Williams Mountain. J. W. Hunt, George W. Orr, Billy Campbell, and W. H. Perry were all cited as being sheepmen in 1876 to 1877. The *Weekly Arizona Miner*, July 20,1877, stated that there were twenty-five thousand sheep grazing here.

Due to the rich grazing lands, others soon followed these men in the 1880s. Gustave Reimer and James May had a partnership in the 1880s. Phillip Hull may have come as early as the late 1870s. Frank Riselda was reported to be here in 1882. Joseph B. Tappan arrived in 1887. He was still involved in the sheep industry in 1889, the year that he assumed the presidency of the Arizona Sheep Breeders and Wool Growers Association. T. Fred Holden settled in Johnson's Canyon, which already had sheep when Captain Johnson brought his sheep there in the late 1870s. One of the largest cattle and sheep ranchers in northern Arizona was Dr. E. B. Perrin. He owned thousands of acres that he had purchased from the railroad and owned the Baca Grant, over a hundred thousand acres in western Yavapai County. He began primarily running cattle but soon turned the grazing land over to mostly sheep. It should be noted that cattle were in the vicinity in the 1870s but by the 1880s were arriving in huge quantities. Cattle were brought in from California for the same reason that sheep were: drought. The cattlemen soon began to crowd the sheepmen by buying sections of railroad grants. They restricted the movement of sheep ranchers and at times took over the sheep rangelands. Frank Rogers reported in 1886 that trouble was anticipated between the cattlemen and sheepmen. In the beginning, the shepherd was "prone to shoot cowboys and cattle on sight."[61]

With the number of sheep in the area, Williams was an important wool and sheep-shipping town. Three years of shipping records of the Atchison, Topeka, and Santa Fe Railroad reveal there was 31 tons of wool in 1885, 62 tons in 1888, and a year later, 77 tons. The doubling of tonnage in three years shows the importance of this area for sheep grazing. Chalender, another nearby community, shipped 30 tons in 1888 and 82 tons in 1889.[62]

Thomas Forsythe McMillan has been credited in some historical sources for having brought the first sheep into the Flagstaff area prior to May 1876. The Pioneers' Society of Northern Arizona recorded in their minutes, dated April 12, 1905, that McMillan arrived in the area no later than May 1876. The Yavapai County records showed that he claimed land and all the water rights in the Little Colorado in February 1876, but as Cline pointed out, the land would have been too small for sheep grazing, suggesting he had other holdings in the area.[63] McMillan herded the sheep in from California, and according to his daughter's interview[64] in 1953, the sheep were originally from Australia. McMillan had been raised by his uncle after the death of his parents, and when he was a young man, he left Tennessee for

---

61 Rufus K. Wyllys, *Arizona: The History of a Frontier State* (Phoenix: Hobson & Herr, 1951), 253.

62 James R. Fuchs, *A History of Williams, Arizona 1876–1951* (Tucson, AZ: University of Arizona Press, 1955), 157–158.

63 Cline, 92.

64 Oral History Interview with Mamie Fleming [includes transcript], March 7, 1953, Northern Arizona University Cline Library.

California and then went to Australia. Returning some time later, his daughter remembered being told that the sheep were brought back with her dad. There has been speculation as to the number of sheep he brought, and it has been suggested that it may have been as many as two hundred thousand, but that number has never actually been determined.

Much of the information gathered about McMillan comes from his daughter, Mamie Fleming. She remembered an incident where he obtained more sheep in a bet made with Hank Lockett, another sheepman at the time. In a poker game, McMillan mentioned he was getting married. Lockett and the other poker players bet him a band a sheep that he would come back single from California. When McMillan arrived back into Flagstaff with his bride in 1888, he went to the Lockett pasture and began to drive the sheep onto his own pasture. A band of sheep could be anywhere from two thousand head to two thousand five hundred head. The total number of sheep McMillan got from Lockett was unknown to the daughter.

McMillan homesteaded in what would be the path of the future Santa Fe Railroad. He had taken "a large tract which included the areas of the city (Flagstaff)...and on out (to) Antelope Valley including the springs."[65] It was part of the southern portion of this land that he sold to Asa Clark, mentioned above. Mrs. Fleming said of her father, "When the Santa Fe Railroad came in, he had to give up his range to the Atlantic and Pacific Railroad. The first

65 Cline, 92.

trip I ever took on a train was when my mother and my two brothers and I went to California to visit my mother's people. We rode on a pass. That was the way the Santa Fe paid my father for giving his grazing land to the railroad."[66]

McMillan also lost land when the Forest Service made large landowners divvy up their lands, according to his daughter. McMillan had built his homestead with a spring for water where the high school was in 1953. At this location, it has been surmised that troops on tour of the country arrived on July 4 and hoisted a flag on a tree, and the name of Flagstaff was born.

Today, his home, a two-story log cabin built in 1886 using wood from his land, is located across from the Museum of Northern Arizona. McMillan had moved several times before he settled here. The home was built before McMillan went to California and got married. His bride was a widow, with a son. Three more children would be born into the union.

Mrs. Fleming continued to talk about her father and his sheep business, stating that their wool was sent by wagons to Prescott, where it was freighted to the Colorado River to be shipped to either the East or West Coast. This would have been prior to the railroad since he settled in the Flagstaff area in 1876. The birth of McMillan's daughter occurred in 1894, a Depression year and thus she was referred to as a "hard times baby." She stated in the 1953 interview that "we had a terrible drought and a national depression during Cleveland's administration, and the

66 Oral History Interview, Fleming, 6.

sheepmen and other men in business had a terrible time. They had no sale of wool, and they hadn't any rain. Their sheep died, and they just had awful times, and my dad really lost an awful lot of sheep."[67]

Fleming also spoke of her father losing sheep to Mexican rustlers from New Mexico when he wintered his sheep near the Indian reservation in a place called McMillan Crack. She said that coyotes got many of the sheep as did wild dogs. McMillan survived the drought of the early 1890s and the lifting of the tariff on wool that made wool prices decline, and when the forest reserves were just in their infancy, he retired and sold the sheep in 1895. Up until his death in 1906, he was active in other business ventures and politics.

One very useful piece of information garnished from Mrs. Fleming's interview is the mention of names that were also listed in Haskett[68] and as reported by The Young and Old Observer[69] [70] [71] These documents help show the number of men involved in the sheep industry from the 1870s to 1890s. Many of these men would continue to operate into the 1900s.

Early written accounts about Arizona are extremely helpful to give a picture of the territory and its econom-

---

67 Ibid., 6.
68 Haskett, 47–48.
69 Frank D. Reeve, ed., "The Sheep Industry in Arizona, 1903," *New Mexico Historical Review*, Vol. 37, no.3.
70 Frank D. Reeve, ed., "The Sheep Industry in Arizona, 1905-1906," *New Mexico Historical Review*, Vol. 37, no. 4.
71 Frank D. Reeve, ed., "The Sheep Industry in Arizona, 1905-1906," *New Mexico Historical Review*, Vol. 38, no. 1.

ic activities. One account from 1877 stated that General Kautz had seen many large bands of sheep in the San Francisco Peaks area. This report by Kautz had to be prior to 1877 and thus shows that the sheep industry was thriving during the mid-1870s.[72]

Frank Hart and his wife, who were friends with McMillan's wife, arrived shortly after the McMillan's marriage. Hart and McMillan partnered in the sheep business until it dissolved with the death of Frank and the rest of the Harts moved back to California around 1888.

Sometime in 1876 or 1877, Walter J. Hill brought in an unknown number of sheep into Arizona, but within twelve years, he was reported to be the largest sheep owner in the territory. The newspaper reported in some years that he averaged a hundred thousand pounds of wool.[73] Sheep usually give an average of about seven pounds of wool, which would place the number of sheep he owned between ten thousand and twenty thousand. Cline[74] estimated that Hill had between twenty thousand and thirty thousand, which is probably high for the average amount of wool per sheep.

An easterner from Ohio, James Houck, came into Arizona in 1870 after he fought in the Civil War. He was considered a tough frontiersman who could take care of

---

72 Richard Hinton, *The Handbook to Arizona: Its Resources, History, Towns, Mines, Ruins, and Scenery* (San Francisco: Payot, Upham & Co., New York: American News Co., 1878), 299.

73 Cited in Cline, 96.

74 Cline, 96. Twenty thousand sheep times an average wool clip of seven pounds per sheep would be 140,000 pounds, almost half more than the 100,000 averaged. Thirty thousand sheep would yield 210,000 pounds.

himself. He did not intend to herd sheep but wanted to try his luck in the gold mines. After four years of searching for his fortune in gold, he took up a job as a military dispatcher and carried mail to out-of-the-way posts. The job was dangerous, and it paid well as a result. He continued in this line of work for three years. His fast horses never let him down, and he was always able to outrun the Apaches. His next adventure was a small trading post near present-day Holbrook, Arizona. Houck, Arizona is named after him. This was Navajo Country, and while they did not want him here, he won their respect with his tough attitude. One incident showed this attitude:

> The trader was sometimes hard-pressed to keep a jump ahead of the light-fingered among his customers. One morning he awoke to find missing from his pasture a fine black mare and colt that he had recently brought. Knowing that it was useless to confront the Indians head-on, he just closed the door of his store and waited, turning away all who showed up to trade. After a few days, one of the leaders came down to find out what was going on. Houck told him that the store would open when his horses were returned. The next morning, the mare and colt were back in his pasture and the door to the trading post was open again.[75]

Nothing has been written about Houck and the sheep that he acquired from the Navajo until his move to Heber, Arizona. Within five years of his marriage, Houck moved to the Heber area, where he and his wife bought the JDH

---

75 Frances Carlson, "James D. Houck: The Sheep King of Cave Creek," *The Journal of Arizona History* 21, 1: 44.

Ranch, grazing both sheep and cattle. Houck also was involved in the Pleasant Valley War between the Grahams and Tewksburys, a war that was about where sheep versus cattle could graze.

Houck concentrated on his sheep business, and it was soon to take a turn in another direction. During this time, the Daggs started to winter their sheep in Paradise Valley, north of present-day Phoenix. The first outfit to drive sheep into the foothills and desert area of the Agua Fria River was Campbell and Francis in 1891. This outfit had between three thousand and five thousand sheep. The winter rains in the lower desert supplied ample grasses for the flocks. The Daggs alone brought thirty thousand sheep into Paradise Valley for shearing and lambing in the spring before they took the flock back to the north for the summer. Shearing could take place earlier in the warmer environment of the desert, and the young lambs could have a better chance of survival. Both the wool clip and the newborns were important to the herder as the wool paid his expenses for the year and the newborns were the owner's profit.

The Daggs were followed by other sheep outfits. Between 1895 and 1910, thousands of sheep were grazing in the low desert in the winter.[76] Carlson stated that by the turn of the century, two hundred thousand sheep were "wintering in the Salt River Valley, most of them spreading out over the desert and foothills north of the Arizona

---

76 Some of this information comes from an unidentified writer who titled his paper "Notes Early History of Livestock in Arizona," Arizona Wool Growers Association, NAU.MS.233, Cline Library. Special Collections and Archives Dept.

Canal to Bloody Basin and from the Agua Fria River to the Verde (River)."[77]

Houck moved south with the sheepmen. After he looked around and saw all the makeshift camps that had been set up for the sheepherders and shearers, being the practical and shrewd businessman he was, he realized that a permanent camp was needed for the shearing and so wool buyers could purchase the wool.[78] His shearing station, Cave Creek Station, was near present-day Cave Creek, just north of Phoenix. Everything he needed was there: a school for his children, a reliable water source, and plenty of surrounding countryside ideal for winter sheep grazing. He also set up a service that would haul mail and passengers between his ranch and Phoenix. His store received supplies on the return trip, and when it was shearing season, he added a burro pack. More buildings were added as the population swelled, and there was a need for feeding and boarding the variety of people who came through. He even hired a Japanese man to grow vegetables for feeding nearly a hundred people daily during the busiest times. With the only liquor license in proximity, he packed a six-shooter, and it was known that when he buckled on his gun belt, the ruckus had better cease. Houck was the first establishment to add a shearing machine run by gasoline.

During the height of his shearing business, he was called "the sheep king of Cave Creek."[79] During the year,

77 Carlson, 47.
78 Ibid., 48.
79 Ibid., 54.

the Houcks made money listing their ranch as a health resort, and they rented tents to those with tuberculosis. He used the stage line in the summer to bring campers to a place with temperatures cooler than Phoenix. The boarding house was used for entertainment during the off-shearing season. Houck knew how to keep his operation going year-round. He became active in politics again, renewing friendships with those he had known while in the state legislature. When mining increased in the area, the Houcks profited, but, darker days were coming.

Drought years saw few sheep come for shearing as the sheep wintered elsewhere. It was too costly for sheepherders to trail their sheep back to the area because the ewes would be stressed, thereby reducing the lamb crop. The Tonto National Forest had restrictions in 1908 on the number of sheep that could graze its land. Soon most of the sheep were grazing far from Houck's operation. Mining also dwindled, and Houck saw a major downturn in his finances. He divorced, and his life ended when he took a lethal dose of poison in 1921.

When sheep wintered in the Salt River Valley, it was reported by the *Coconino Weekly* Sun, March 3, 1892, that 80 percent of the lambs were saved because the weather was obviously warmer. In addition, the lambs were born in March, so the northern sheepmen had a three-month advantage of growth in the young wethers when they sold in the Flagstaff area. In the north, lambs were not born until June.

Early newspapers are another good source of information about what was happening during various years. It would be a monumental task to shift through all the territorial newspapers and those that continued or started after statehood. Looking through several years of the *Flagstaff Sun Democrat* and *Coconino Weekly Sun*, bits and pieces of information help tell the story of the industry. Names that have not shown up in other sources come to light. An interesting section to browse was "Local Brevities." Winter rains during 1891–92 had been good were reported in the *Coconino Weekly Sun*, March 17, 1892, and the grasses at Padre Canyon and Canyon Diablo were a month ahead of last year. In the year 1897, the paper cited sheepherders that were known to the author and what they were doing: mainly visiting or having visitors and some news about the sheep business.

For example, the July 22, 1897 edition of *The Flagstaff-Sun Democrat*, stated that Thomas A. Sayer had sold two thousand head of fine wethers to Jerry Woodbridge, and Woodbridge was going to drive them to New Mexico for grazing, but that same issue stated, "Lucien G. Smith, one of the staunch wool growers of northern Arizona, was in the city Tuesday." Lucien Smith is not one of the names on the list given by Haskett. If he was a staunch woolgrower as the newspaper claimed, why did Haskett not list him? The same statement was made in the newspaper for James W. Cart being one of the most successful woolgrowers of Navajo County; his name is on Haskett's list.

Newspapers also record the dissolution of companies, which helped explain why a sheep company just disappears from the record. In the July 15, 1897, and several later editions, the *The Flagstaff-Sun Democrat* reported a notice of dissolution for Dent and Sawyer and indicated any money owed to them should be paid to Sawyer.

These newspapers helped provide a picture of the problems that both the cattlemen and sheepmen had to face. The cattlemen were working with the legislature to change territorial laws to protect the grazers of the territory. They thought that the local sheepmen had "a right to a fair share of the public domain and the cattlemen will never raise an objection," was reported in the *Coconino Weekly Sun*, February 25, 1892, but the article continued, the cattlemen were opposed to large flocks being driven into their ranges from outside the territory.

One early but short-lived sheep pioneer was John Elden. He arrived into the area between 1875 and 1877. His daughter told the *Coconino Sun* that it in 1875[80] he brought his wife back from California. They probably herded a flock out of drought-stricken California. A log cabin was built at the base of what would become known as Elden Mountain, and the couple raised three children. The youngest was a son who was tragically killed while tending sheep. Shortly after his death, the Eldens sold out and left the area.

While there is no definitive date for the start of the Frisco Sheep Company, it was here prior to the railroad

---

80 Cline cites 1877 as the date, 98.

arriving in Flagstaff in 1883. The first owner was Charlie Schultz, who ran the sheep through Schultz Pass. He sold the outfit to Mr. M. I. Powers in 1910. Mr. Powers bought the Brookbanks Ranch with its 320 patented acres on Elden Mountain sometime after buying the Frisco Sheep Outfit and grazed his four thousand sheep there. The September 2, 1927 issue of the *Coconino Sun*, noted he sold the sheep, other livestock, the ranch, and rangelands in and northeast of Schulz Pass, along with his Flagstaff residence on Leroux Street, to Mr. Ysi Otondo of Winslow. Ysi Otondo was the brother-in-law of Mr. Powers' foreman, Cruz Eraso.

At the time of purchase, Ysi Otondo had no sheep of his own as he had disposed of his interest sometime in 1926 when he went to the Mayo Brothers (now Mayo Clinic) for an operation. He could not stay away from his choice of profession and thus bought back into the sheep business with this purchase of the Frisco Sheep Company. Ysi Otondo had been a partner for the past eighteen years with the large sheep outfit operating under the name of Sawyer & Otondo in Winslow.

Herding sheep from New Mexico, Felix Scott brought his sheep into the Little Colorado River Valley in 1874. It is uncertain as to the number of sheep Scott brought with him. Like Scott, many names were given in the early history of the state, but the only information may be the sheepherder's name and sometimes the number of sheep and where the sheep grazed. This information shows that many men were involved in the beginning and each helped make the sheep industry.

Two of the three Daggs, J. F., and W. A., followed others into the northern portion of the state, settling in Apache County. A great deal of information has been written about the Daggs, and, in part, this was a result of their involvement in the Pleasant Valley War. J. F. and W. A. Daggs, known as Frank and Will, had purchased their sheep from a well-known Basque sheep breeder in California, Domingo Amestoy. J. F. grazed his sheep on Anderson Mesa at Ashurst Run. Will grazed in the area of Silver Creek, a tributary of the Little Colorado, in the White Mountains. They purchased sheep from the Hunings of New Mexico and from a sheepman in the Holbrook area. To obtain these sheep, they traded their ranch on Silver Creek to buy three thousand wethers from the Hunings of New Mexico and moved their operation to Chavez Pass, where water was readily available at a well. The wethers were each sold for two dollars. When the third Daggs joined them, they ran their sheep under the name of Daggs Brothers & Company and were headquartered in Flagstaff until 1890. Most of their sheep were on shares with sheepmen like Alfred and Frank Beasley, Jose P. Chavez, Paulitta, Manuel Guterias, Ike Wheeler, Bill Yeager, Manuel Alaguar, Oscar Varden, and Rube Olander.[81] The Daggs received a percentage of the lambs and wool from these shares. By 1885, it was reported that they had as many as one hundred and fifty thousand sheep.[82]

---

81 Not all of those listed are also found on Haskett's list.

82 Maurice Kildare, "Jose Chavez: The Man Who Refused to Die," *Real West* (March 1968): 41.

The sheep in Arizona at the time of the Daggs were considered of "a poor lot reduced to mere runts by inter-breeding."[83] Wentworth stated, however, that the Arizona "sheep had several valuable characteristics. During the centuries in which many assumed that they had been de-generating, they had learned to rustle for food and water, resist storms, and elude wild animals. They provided Arizona sheepmen with an excellent foundation herd thoroughly adapted to the environment and climate."[84] Thus, enter the Daggs.

The Daggs are credited with improving the breed in Ar-izona when they purchased sixty-six pure bred Spanish and French merino rams from the Vermont Merino Sheep Breed-er's Association. The sheep had a price tag of between $100 and $600 each, a great deal of money in 1882! These rams were placed with the well-bred ewes from California, and by 1886, the Daggs started to see a profit on their purchase. The thousand rams produced were sold to other Arizona breeders from six to twelve dollars each. The Daggs adver-tised statewide and "developed show flocks whose rams swept the premium lists at the state[85] fairs in Albuquerque from 1884 to 1886."[86] The wool from these sheep averaged eight pounds per sheep, and from 1884 to 1887, the price of wool was estimated to be fifteen cents per pound.[87]

---

83 Hamilton, 285.

84 Wentworth, 250.

85 This actually should be territories; neither New Mexico nor Arizona was a state until 1912.

86 Wentworth, 250.

87 Robert Clark Euler, *A Half Century of Economic Development in Northern Arizona 1863–1912* (Flagstaff: Arizona State College Master's Thesis, 1967), 61.

But that was not all the Daggs would be known for in the territory. The Daggs would be part of an important organization formed to promote and help protect the interest of the sheepmen. As for the rest of the story of the Daggs, they would operate as Daggs Brothers & Company until 1890, when they liquidated the company and each would have their own sheep outfit. With the economic downturn that took place in 1893, no one risked purchasing more high-bred rams to continue the fine-wooled sheep.[88]

Harry Fulton also made efforts to improve the breed of sheep in Arizona in the 1880s. Fulton arrived in Arizona in 1876, making his headquarters in Prescott, where he handled sheep on shares. After purchasing a large number of sheep in 1881, he moved the headquarters to Flagstaff. Herding on the Colorado Plateau and in the San Francisco Mountains, he had between two thousand and nine thousand head of sheep.[89] Fulton tried to improve on a combination wool and mutton type. He crossed the Rambouillet with the English down breeds. Fulton got a good mutton sheep, but these animals were "unsuited for breeding purposes on the Arizona ranges. Like that of others, his search for a dual-purpose sheep adapted to the Southwest was a failure."[90]

Two other sheep outfits operating in the Flagstaff area during the 1880s were the Moritz Lake Sheep Company and the Asa Sheep Company. Joseph Moritz arrived in Flagstaff in 1883 and bought an interest in his cousin's outfit. The

88 Wentworth, 1958, 251.
89 *Portrait and Biographical Record of Arizona*, 652.
90 Haskett, 25.

number of sheep his cousin, J. S. Miller, was running in the area surrounding Flagstaff is unknown. Moritz acquired sheep from Perrin and Eldon. In 1900, the Forest Service refused to grant Moritz grazing rights, so his sheep began to die. When he sold to the Babbitt brothers, he had five thousand sheep. The Asa Sheep Company began with the leasing of three thousand shares from the Daggs in 1886. Alfred Beasley was the original operator of the company. He received half of the wool and half of the newborns with the rest going to the Daggs. He sold his 7,500 sheep to the Babbitt brothers in 1900, making $75,000.[91]

Ramon Aso was a Spanish-born Basque, well acquainted with sheepherding. He had arrived into the United States on the day San Francisco suffered both an earthquake and fire, 1906. He was nineteen years old. For several years, he worked for the Lockett Sheep Company. On the recommendation of Alfred Beasley, he ran sheep for the Babbitts, where he was able to acquire a fourth interest in the sheep. Sometime later, Aso withdrew from Babbitt's company. In 1927, he bought the Howard Sheep Company and made his headquarters near Howard Lake Ranch.[92]

The Arizona Sheep Breeders and Wool Growers Association

One of the biggest threats to the sheepmen from the beginning were the cattlemen. As cattle were pushing into

---

91 Euler, 66.
92 Arizona Wool Growers Association, NAU.MS.233, Cline Library Special Collections and Archives Dept. Series 1.2.40.

the northern ranges of Arizona, the sheep were pushed aside. Conflict developed, the most famous being the Pleasant Valley War. With growing pressure, the sheep ranchers organized the Arizona Sheep Breeders and Wool Growers Association in 1886 at a conference in Flagstaff. Its purpose was "to promote the breeding and use of purebred rams, to arrange for the annual rodeo conducted each year for the purpose of going through man's herd and removing the stray sheep and returning them to their respective owners, to agree on a uniform wage scale for herders and shearers, and to assist the industry generally on all matters of common interest."[93] Sheepmen benefitted from joining the organization because there were advantages to numbers of men stating their opinions and forming a consensus that the association could present on the many issues facing the wool raisers.

To understand the importance of an organization such as the Arizona Sheep Breeders and Wool Growers Association, it is necessary to understand the politics of the period. Arizona was open rangeland through the 1890s. During the 1890s, there were signs of overgrazing. With the Forest Reserve Act of 1891, the president of the United States had the right to set aside forest reserves from the public domain. President Harrison was the first to use the act, followed by Presidents Cleveland and McKinley. The law was designed to protect the remaining stands of valuable timber and especially the timber found at the headwaters of streams to prevent devastating floods and

---

93 Haskett, 27.

to ensure there would be a water supply for the arid western lands through the summer.

Two years of drought began the spring of 1891 and had dire consequences. Grasses that once were as high as the belly of a horse were gone, as were the water holes. It was estimated that too many cattle, sheep, and horses were grazing the land, even in good years. It was becoming obvious to many that sheep, goats, cattle, and horses were overgrazing. When the rains came in 1892, those areas where the roots survived grew again, but there were many signs of overgrazing, so the vegetation never recovered, and the heavy rains produced gullies. Wildlife was affected. During these years, prices were high, animals were shipped off to markets, and overgrazing continued. Then the economic downturn hit in 1893, with wool prices low because the tariff on wool was lifted. Cattle and sheep prices both decreased. In 1895, prices increased, and those that had survived either added to their flocks or sold out, like McMillan of Flagstaff.

On August 17, 1898, President McKinley established the San Francisco Mountain Forest Reserve, now the Coconino National Forest, but because of sections of land given to the railroad as it was built across the country, the reserve was not one solid piece of land. The railroad and other groups owning land in the reserve wanted to trade for more lucrative land and to ban sheep grazing; the railroad stated the purpose for trading the land was to consolidate it. Strong opposition built across the country against the sheepmen, and much of it was clandestine.

To counter these attacks, the Arizona Sheep Breeders and Wool Growers Association revamped as the Arizona Wool Growers Association (AWGA) in 1898. Heated arguments can be found in the newspapers about sheep and the reserves. *The Coconino Sun, August 31, 1899* stated why the AWGA was for President McKinley and stated that investigations "will show that sheep grazing under proper rules and restrictions can be made the strongest safeguard for the forest reserve of Arizona." The article continued with a statement from AWGA that said it was a fallacy that sheep "injured the forest growth and interfered with the water supply of the southern portion of the territory." The sheepmen did not intend to overgraze these lands any more than the cattlemen did, but this ban would mean the demise of the sheep industry within the state as they would have no grazing land.

With the drought of the early 1890s, conservationists accurately assessed the damage to the forest. The coffers of the cattlemen were empty, and they could not gather support to fight back. Rangers were sent to enforce the grazing rules that each stockman must have a permit to graze on public domain.[94]

Gifford Pinchot, a forester, and Fredrick V. Colville, a specialist of the Bureau of Plant Industry, investigated. Though not released by the secretary of the interior, it was surmised that grazing sheep in the forest was favorable. In 1901, the sheep were allowed to graze the forest reserves. Powerful allies prevailed on President Roosevelt, however,

---

94 Peplow, 154–155.

to have sheep banned, but once again, the strong influence of Gifford Pinchot, who was a longtime friend of E. S. Gosney, the president of the AWGA, the ban was lifted.

Undergoing growing pains, the forest reserves were finally placed under the Department of Agriculture. Permits were issued for grazing a specific number of sheep on a definite allotment from June 1 to November 1. Starting in 1905, fees were charged per animal. The animals could move to these summer grazing lands over driveways established by a committee of one cattleman, one sheepman, and a forester.

As president of the AWGA, E. S. Gosney[95] worked diligently to secure privileges so the sheep could graze on government lands under regulations that were agreeable to both the sheepmen and the government. The organization secured the allotments for each sheep outfit, helped make the trails for movement of the sheep each spring and fall, and protected the range. There were other functions as the years went on, such as keeping the sheepmen informed of legislation that could affect them. The organization still exists today in Arizona.

Each year, the AWGA would meet, and one of the items on the agenda was to meet with the superintendent of the forest reserve to adjust the grazing permits. This was not unique to Arizona; any state where sheep were allowed to graze on government lands like forests would have an association that met to adjust the grazing permits. Many of the sheepmen realized that without

---

95 Gosney was president of the AWGA from 1898 to 1907 and 1908 to 1909.

the state organization, they could have lost their right to graze sheep on these government lands, as that was the idea of the Department of the Interior.

The Railroad Brings Changes

The advent of the railroad was one of the most important events to encourage and promote the sheep industry in the Arizona territory. This is not the story of the building of the railroad in Arizona, but the railroad surely played a role in the sheep industry. The Atlantic and Pacific Railroad, also known as the Santa Fe Railroad, allowed wool and meat to ship out of the area and made bringing sheep into the territory easier. The northern route of the railway across Arizona was completed in 1883. Before the railroad, wool and sheep meat was sent by oxen carts to New Mexico, where the railroad would then haul to eastern markets. The alternative was to freight to the Colorado River and then onward to a port in Mexico, then head to the eastern United States by going around the tip of South America.

The railroad also created the demise of freighters such as Tully, Ochoa, and Company and Lord and Williams, mentioned before. These two companies were also large sheep ranchers in southern Arizona and brought sheep into the area in the early 1870s. It can only be speculated that when their freight businesses went bankrupt, their sheep businesses did also as the two outfits are not mentioned again.

When the government approved building the railroad at the company's expense, the government gave alternate sections of land in thirty-mile-wide increments on both sides of the track for the railroad to sell to help defray the cost of the railway. If any land was already owned in this right-of-way, the railroad was given land ten miles north or south of the track instead. Stockmen, lumbermen, prospective settlers, and farmers quickly bought these sections of land, but the settlers and farmers soon sold their lands because the inadequate rainfall didn't allow crops to grow. The land quickly reverted to grasses—idle land to graze livestock, both cattle and sheep.

When the railroad was completed, the headquarters of the wool growers switched from the Silver Creek area south of Holbrook to Flagstaff and along the line of the railroad. Prescott was already seeing a shift away to the north and the Bill Williams Mountain. Sheepmen began to call Flagstaff home in the 1880s, and they were instrumental in the sheep industry. They included the Riordans, E. S. Gosney, John Noble, the Locketts, D. M. Francis, J. Reimer, the Daggs brothers, and Dr. E. B. Perrin. In the mid-1890s, Gosney owned three ranches and had between six and twelve thousand sheep along with an unreported number of cattle.[96] The Bly Brothers and Sawyer and Otondo shifted their operations to Winslow. Shearing pens and corrals would be constructed to make shipping wool and animals easier.[97]

Shearing operations were moved along the railway and the amount of wool shipped eastward was reported

---

96 *Portrait and Biographical Record of Arizona* (Chicago: Chapman Publishing Co., 1901), 661
97 Wentworth, 251.

by Governor Tritle in 1883 to be approximately 2.4 million pounds, valued at over $500,000 (approximately $13 million in 2015 dollars) as the average price of a pound of wool was put at twenty-two cents. With the increase in the grade of sheep, more and better wool was the result. The governor wrote two years later that the shearing of merinos brought $3.50 a piece, the improved Mexican strains were selling between $1.50 and $2 a sheep, and wool was selling between sixteen and twenty-one cents per pound in the East, obviously less than two years prior. These prices did not include the freight charges for the wool being shipped between Flagstaff and Philadelphia, which was five cents a pound. A railcar could hold approximately twenty thousand pounds of wool and cost the wool growers $1,000 per car.

Loading sheep for shipment to Salt River Valley, circa early 50s.
Photograph courtesy of Frank Jr. and Marcie Erramuzpe.

With the shipment of sheep to eastern markets, especially Kansas City, new laws were enacted to protect the livestock. Before 1875, the Cruelty to Animals Act stated that livestock had to be unloaded, watered, and fed every twenty-eight hours. There were two main problems with this, however. First, all livestock were considered to have the same needs for food and drink. The entrusted agent was responsible for unloading, watering, and feeding of the animals. Otherwise, the sheepmen themselves would have to travel with the sheep and would be responsible for these activities. The sheepmen needed the sheep to arrive at market in the best possible condition to secure the best price for the animals based on weight. If the animals were mistreated while they were being loaded or unloaded or were taken off the train into mud, the animals would not eat or drink. Sometimes the animal just would not eat anyway. The second problem was that these twenty-eight-hour interval forced stops were paid for by the stockmen. If there was a delay, the cost was passed onto the stockmen as well. The Old Observer wrote that sheep "may be shipped considerably longer, without injury, than cattle may be."[98] The sheepmen wanted the law repealed or at least revised to state the needs of the different types of livestock shipped.

The actual number of sheep in the territory during the 1890s may have been as high as seven hundred thousand, but, as stated earlier, the numbers varied considerable between researchers writing in the 1940s and 1950s.

---

98 Reeve, October 1963, 341.

Wentworth stated the number, according to the US Census, to be 102,427. Researching the 1890 US Census reveals it recorded 924,761 sheep within the territory, a 215 percent change from the 1880s census. Wentworth[99] concluded that the US Census did not include the large flocks that the Navajo and Hopi raised on their reservations. The government's report of 1892,[100] which stated that non-Indians ran 803 sheep in 1870, 76,524 in 1880 and 698,404 in 1890, confirmed this. This report does not resolve a discrepancy in the data for the number of sheep in the territory for 1890.

An interesting statement in the report was that "sheep husbandry is the leading and most profitable agricultural pursuits of its people, and the indications are that this will continue the leading livestock business."[101] Arizona was considered an ideal area to graze sheep in the 1890s, and many government reports attest to this fact.[102] This was evident in the number of flocks that were grazed here from New Mexico, Colorado, Utah, Nevada, and California. Wentworth[103] believed that the advertising by the railroad did more to bring in sheep than farmers and homesteaders. At this time, there were no barriers to grazing, and the large tracts of open grasses were incentives for wool growers, but this also led to problems with the cattlemen.

---

99 Reeve, July 1963, 253.

100 U.S. Department of Agriculture, Bureau of Animal Industry, Part II, 935.

101 Ibid., 935.

102 Ibid. 935–945. There are many reports imbedded in this document to substantiate the importance of the sheep industry in Arizona.

103 Wentworth, 254.

Clashes with Cattlemen

Peplow[104] stated that sheep and cattle both grazed what was considered inexhaustible grasses during the 1880s and 1890s. Flocks of sheep were grazed in the San Francisco, Mogollon, and White Mountains area during the summer and then trailed to warmer grazing areas further south. The cattlemen began to move their herds into these grasslands of the north. They also needed the railroad to ship their cattle to eastern markets. When the large cattle outfits began to spring up, such as the Aztec Land and Cattle Company[105], their sheer size could put pressure on the sheepmen and push them out. Haskett stated, "At that time it was the unwritten law of the range that he who first watered his stock at a stream, spring, or waterhole had the prior or exclusive right to its use thereafter together with such adjoining range lands as he could use."[106]

Even though sheep had arrived with the Coronado Expedition, the cattlemen believed that they had established their claims first, and they did not want the sheep grazing on what they considered their territory. It was also believed that the sheep were harmful to the environment, but even before large herds of sheep were in Arizona, the land was overgrazed.

A line had been drawn in the sand, the so-called deadline that sheep should not go beyond, the Mogollon Rim.

---

104 Peplow, 151.
105 Wentworth, 253. Wentworth states that the Aztec Land and Cattle Company "was the Hashknife Outfit as named from it brand." Hashknife Outfit will be used hereafter unless found in a quote.
106 Haskett, 26.

The Tonto Basin had prime grasslands. Two large cattle companies put obstacles where the sheep could graze. "The A One Bar cattle company at Flagstaff blocked them from extending north. The Hashknife or Aztec Land and Cattle Company formed an insurmountable barrier against ranging east across Apache County into New Mexico."[107]

The Hashknife Outfit was formed in 1884. The company was made up of wealthy businessmen from the east and made shrewd business decisions, especially when they purchased one million alternating sections of land along the Santa Fe Railroad. Because of the checkerboard pattern, they had tied up more than two million acres along the railroad and could control usage and access to those areas they did not own even though their land had not been fenced.

While the Pleasant Valley War brought the most deaths for sheep, sheepmen, and cattlemen, it was not the first battle between the two groups. There had been many incidents of sheep being driven over cliffs after their herder was tied up or killed. Cowboys from the Hashknife Outfit were notorious for using a variety of methods to keep settlers and sheep off their land, even to cross over to open sections still open to settlement. "They had been running herders off the range for some time. Herders on the Little Colorado River lost over four thousand head when irate cattlemen ran their sheep into the river. North of Flagstaff, ten bands (25,000) of sheep, near the San Francisco Peaks,

---

107 Kildare, 41.

were mangled when a herd of horses stampeded through their midst."[108]

One more story illustrates the length the cattlemen went regarding sheepherders. In 1883, a sheepherder had all his sheep run off. He was to be hung, but it was decided to notch his ears instead. The herder lived, and two years later sued for $10,000, won and then disappeared.[109]

The story of the Pleasant Valley War begins when John, Jim, and Ed Tewksbury took eight thousand of Daggs' sheep on shares to the Tonto Basin to supposedly provoke the Grahams, who were grazing their cattle. The Daggs were very well aware of the feud between the Tewksburys and the Grahams. As the story goes, the Daggs suggested the Tewksburys take some of their sheep into the Tonto Basin, and "it would offer not only the opportunity for the Tewksburys to make a profit from their land, but would also allow them further retaliation against the Grahams by running the cattle business out of the area below the Mogollon Rim."[110] Both the Tewksburys and Grahams were cattle rustlers, and bad blood had been escalating. They had disagreed over cattle they had stolen together.

At first the Grahams did not fear the arrival of the sheep, but they soon learned more were on the way. The sheepmen were not going to use the choice grassland. The Daggs were warned that a battle was about to take place,

---

108 H. Henry Sheffer and Sharyn R. Alger, *The Pleasant Valley War Cattle and Sheep Don't Mix* (Apache Junction, AZ: Norseman Publications, 1994), 6.
109 Ibid., 6.
110 Ibid., 7.

and they gathered forty men to herd the sheep out of the Tonto Basin. Most of these men were not fighters as some just worked for the Daggs. Jose P. Chavez was one of the forty men accompanying the Daggs. Frank Daggs figured that since they were heading to the basin to remove the sheep, a conflict could be avoided, but he was wrong. The cattlemen had agreed earlier in their planning against the sheep and their herders that no sheep would ever make it back up on the rim. After many days of trying to move the sheep out of the basin and up onto the rim, constantly being attacked by the cattlemen, only a thousand sheep made it out. Besides the seven thousand sheep, cattlemen and sheepmen were also dead. Jose Chavez had been wounded earlier in the fight but managed to crawl up the rim after four weeks and hailed some sheepherders.[111] Since the details of the actual war in the Tonto Basin are not germane to this history per se, it is enough to say that it was the bloodiest of the sheep–cattle clashes.

More clashes and deaths can be found between the two groups into the early 1920s. Two herders were killed by cattlemen as late as 1920 near Williams. In the *Arizona Republic*, August 13, 1920, issue a front-page story reported that the two deaths stemmed from the Mexican herders trespassing on the land belonging to the two accused of killing, Johnson and Robinson. Johnson had two weeks prior sworn out a warrant against the two herders for trespassing. Johnson and Robinson said they were acting in self-defense as the herders shot first.

---

111 Kildare, 78.

At the beginning of 1890, the sheep industry looked healthy. There were 698,401 sheep reported that year. It was estimated that 35,000 wethers were shipped out that year weighing between 95 to 115 pounds going to both Pacific and East Coast markets. In addition, 9 million pounds of wool shipped mostly to the East Coast. The average fleece weight was seven pounds, but the sheep industry and the nation's economy were hit hard during the 1893 depression. For the sheep industry, it meant a decrease in the price of wool from twenty-two to twenty-seven cents per pound to between seven and thirteen cents a pound. The number of sheep had also decreased from 698,401 sheep in 1890 to 397,460 in 1894.[112] Depressed prices for wages for herders and shearers, wool prices, and animals continued until 1896. Many outfits sold out, and many weathered this storm, just like so many of the sheepmen. It was reported that for every one sheepman to go under, there were ten cattlemen. The sheep industry recovered by the beginning of 1900 with 861,761 sheep reported in the state.[113]

The Young Observer: Early Firsthand Accounts of the Sheep Industry

Early reports about the sheep industry in the Arizona territory came from a person writing under the name "The Young Observer." In at least two bulletins, one dated 1903 and the other 1906, a description of the land,

---

112 Wentworth, 254.
113 Haskett, 33–35.

names associated with the sheepmen, and the number of sheep were given. Problems faced by the sheepmen were discussed as well. Some ranchers had a great deal of information written about them, but others only had their names listed and the number of sheep they owned. Much of this information has already been related in this narrative, but the picture of everyday activities of the sheepmen and their concerns for forest usage is new information.

> "It is just this way, my friend, I contend that matters are radically wrong when one man in charge of the government forest reserves holds in his hand the power to either let us continue on our peaceable and prosperous way, or by his edict put us off our accustomed grazing grounds and thereby ruin us. It is bad for the man thus placed in such arbitrary power, as it holds up to him inducements for bribery, favoritism, and corruption. It is correspondingly bad for the sheepman, as he is continually "walking on thin ice," he has no assurance will hold up longer than the present. From a wool grower's and a mutton raiser's point of view, it is a DECIDEDLY WRONG AND DANGEROUS condition of affairs."[114]

Another sheepman told The Young Observer that he had made improvements on his ranch, which was on forest reserve land, that amounted to $3,000. If he was denied grazing rights, these improvements would be worthless, and his twenty-nine years in the business would be lost. The man stated that with few areas, Arizona was a grazing and mining territory. The Young Observer stated that he

---

114 Reeve, July 1963, 247.

believed that a dam built in the Tonto Basin would add agricultural land, but even with a dam, the majority of the land was only suitable for the grazing of sheep. "Sheep do better and BRING SURER DIVIDENDS on the investment than do cattle in Arizona," he concluded.[115] The Young Observer wrote that sheepmen were more prosperous than the cattlemen were; for every one sheepman going bankrupt, there would be a dozen or more who would prosper in the business. He further stated that for every successful cattleman, there were a dozen or more who would go under. Unfortunately, his statement must be taken with a grain of salt as he does not back up his statements with fact.

So, while his statements about the economic well-being of the sheepmen over the cattlemen may be exaggerated, he describes what occurred throughout the year, especially the movement of the sheep from the mountains to the desert valley area to take advantage of the grasses and wildflowers brought on by the winter rains, the shearing of the animals, and the fattening of the animals to be sold for mutton. If the rains don't come in the desert valleys, there are just fewer lambs as the northern temperatures are not conducive to birthing lambs. The Young Observer reported that for 1903, the grasses and wildflowers were in abundance all the way from Phoenix to Mayer and that it was a good year for the sheepmen. However, the previous winter had seen hardships, with record snowfalls and the isolation of sheepherders being with the sheep for long periods of time.

---

115 Ibid., 247.

The information on some of the early sheepmen—men like Hugh Campbell and his brother, Colin Campbell; Lockett; Woods; Perkins; Sawyer; Hart; and Bly—was garnished from this early report. Detailed information was given for some, like Colin Campbell having between one thousand and one thousand two hundred head of sheep and that he had the best wool because after twenty-seven years in the business, he had a good breed of sheep. Colin Campbell had his headquarters at Ash Fork. He was well regarded by other sheepmen because of his breeding of sheep and the quality of wool from his sheep. Other sheepmen used his rams to help improve their stock. Some of the other mentioned men stated that they were involved in other businesses and were successful businessmen.

Information on Hugh Campbell comes from the *Coconino Sun*, July 12, 1923. Born in 1862 in Nova Scotia to Scottish parents, he set out as a young boy on his own, first lumbering in Wisconsin and finally coming to Arizona when he was twenty. He was a tie contractor for the Atlantic and Pacific Railroad Company. It is not known why he went into the livestock business in 1885. The Campbell Francis Company, where he partnered with Dan M. Francis, became one of the largest sheep outfits in the West by the time of Campbell's death in 1923. Hugh married in 1893, and the couple had two children. His son, Dannie, was a successful sheepman himself. Hugh served as president of the AWGA from 1910 until 1923, resigning just a few days prior to his death.

The Young Observer indicated that the quantity and quality of the wool per sheep was based on the location of a herd. The three areas where sheep are shorn are Phoenix (between five to nine pounds per sheep), Winslow (eight to ten pounds per sheep), and Ash Fork and Seligman (seven to eight pounds per sheep). Sheep that were run in the mountain areas in the winter had a whiter, cleaner fleece with less shrinkage compared to those run in the plains areas.

The Young Observer also wrote two years later concerning Arizona sheep, but he now called himself The Old Observer. This time, he rode the train from an unknown location in California to at least Seligman, describing in detail their path and relating the hardships that the sheep and men must endure to reach the area near Seligman, an area he said was the farthest west sheep could be grazed in the state. In this area, at least at the time he traveled with the sheep, there was an abundance of gamma grass. "This feed is found everywhere—in the open and among the cacti, chemise and mesquite, and the sage brush growth on the higher hillsides and mesas."[116] The only drawback to this area was the lack of water, and to solve this problem, sheepmen and cattlemen alike built storage basins and water holes within a distance of the mesas and far enough from another storage location to not interfere with each other for the capture of the water.

Mr. Edgar T. Smith, who began his sheep business in 1893, was given much attention in this report. Smith had

---

116 Reeve, October 1963, 326.

between eighteen thousand and twenty thousand fine medium merinos. His one band of sheep lambed on grass feed that was reported to be six inches tall in an area a hundred miles to the south of Seligman. The quality of Smith's sheep was traced to him buying from other sheepmen, and their locations were given. It seemed that the sheepmen were proud of the quality of sheep that they raised as this was not the first time the sheep lineage was given in great detail.[117]

From The Old Observer, another sheepherder comes to light. The Howard Sheep Company had been incorporated sometime in 1893 or 1894, but Mr. Howard had been in the sheep business since 1881. Mr. Howard told The Old Observer that he averaged no more than twenty-five thousand head of sheep, and by the time the mutton is sold, the herd would only be fifteen thousand. Howard calculated that it cost him between seventy-five to eighty cents per sheep to run them yearly as he did not trail his sheep southward in the winter. To be able to pay his expenses, he needed to have twelve cents per pound for the wool.

In 1903, The Young Observer stated that herders, those who actually spent day in and day out with the sheep, came from Spain and were paid $30 a month plus board.[118] By this time he was writing in 1905 or 1906, and the herders were both Mexicans and French Basque with wage stated to be between $30 and $35 a month. The food for the herder was "'no canned stuff,' but a sufficiency of potatoes, bacon, flour, oatmeal, sugar, and coffee with

117 Reeve, October 1963, 330–331.
118 Reeve, July 1963, 250.

the usual condiments."[119] Of course they also ate mutton from the occasional sheep they butchered. In his description of the herders, it's important to note that that Basque are mentioned as herders for the first time, and they will be addressed presently.

Many more sheep owners are mentioned in the Young and Old Observer reports, but most of it is very brief[120] and does not add to the story except to indicate the number of owners during the beginning of the 1900s. Those in Coconino County included Mudersbach, Henderson and Wolfolk, E. B. Newman, Henry Yaeger, Melbourne, and Beasley. These names are listed by Haskett also. Once again, these names being on both lists substantiate each other in their reliability for those in the industry. Sometimes other tidbits of information can be found about the men as other sources such as newspapers are scoured.

It is important to note that the sheepmen kept abreast of the politics of the time. In the early to mid-1890s, there was discussion of Arizona being once again joined to New Mexico, becoming one state. While many people within the territory objected to joining with New Mexico to get statehood, most of the sheepmen were against the proposal. With fewer sheepmen and sheep in Arizona, they felt that their interests would be overlooked in favor of New Mexico's. Then there was Mr. Reed, who raised sheep in the Ash Fork area. He had been grazing the animals since

---

119 Reeve, October 1963, 332.
120 The information could be the number of sheep each man owned, when he got in the business and grazed his sheep.

1892. He told The Old Observer that he believed that the ranges were being overgrazed and believed that a solution was a reserve where only so many sheep or cattle were permitted, curtailing the overgrazing.[121] In a later edition, Mr. Yaeger tells The Old Observer that he believed the forest reserve was a good thing because it eliminated bickering over grazing land and assigned grazing to allotments.[122]

By 1905, a highly contagious skin disease called scabies was becoming a nuisance in the territory. The Bureau of Animal Industry of the United States Department of Agriculture was directed to determine a way to eradicate the disease. A group of trained veterinarians and others were sent to stamp out the disease. The territorial legislation created the Arizona Sheep Sanitary Commission in 1921 to help the federal government eradicate the disease. Sheep from every part of the state were inspected, and any with the disease were given a treatment, which was making the sheep swim "through a long vat filled with a liquid solution of nicotine or a lime and sulphur preparation of sufficient strength to kill the causative agent of the disease—a small mite, hardly visible to the naked eye."[123]

Eradicating rodents and poisoning or trapping predatory animals took a toll on the flocks and was the Department of Agriculture's responsibility. Eliminating these animals

---

121 Reeve, October 1963, 339.
122 Reeve, January 1964, 56.
123 Haskett, 45.

helped the sheep raiser, but the government would still attempt to increase grazing fees to the point that many herders were put out of business in the 1920s. The AWGA would expound upon the benefit of the wool growers for wool that was shipped and how this benefitted the state's economy in issues of the *Arizona Gazette*, February 11, 13, 18, 19 and 21, 1920. The organization listed benefits to the state's economy and local governments through taxes on homes, employing herders and shearers, suppliers for the sheep camps, and payment to the local farmers for use of their fields. In various files of the AWGA letters can be found that showed the organization continued the fight as other problems arose where a consolidated front was needed.

CHAPTER THREE

STATEHOOD TO THE PRESENT: FAMILIES' STORIES

Many families raised sheep prior to statehood and continued once the state was established in 1912. Information for each family came from a variety of sources, such as newspapers, magazines, biographical sketches written in the 1960s, other early written histories, and personal interviews with those who were in the sheep business or family members who remembered growing up in a sheep rancher's family. There was a considerable amount of data in the AWGA files, which helped draw an accurate picture of the early history of the sheep industry. It was learned during interviews with family members that previous information written about the family or family member was not always accurate. Where possible, the author asked families to verify information in the historical documents to ensure an accurate picture of each family.

Sanford W. Jaques[124]

The White Mountains can be thankful that Sanford W. Jaques called the area home, as he not only had a large sheep and cattle business in the area but also was

---

124 Paul W. Pollock, "Sanford W. Jaques," *American Biographical Encyclopedia* (Phoenix: Paul W. Pollock, 1967), 80.

instrumental in bringing electricity to the mountain communities. Jaques was born in 1884 in St. Johns. His father died when he was four years old, and his step-father introduced him to sheep. When he was fourteen, he took over the running of the sheep for his stepfather. The outfit was the Scott Sheep Company. Scott ran his sheep on what is now the bottom of Show Low Lake. Jaques was nicknamed "Sant." In 1906, Jaques bought his own sheep herd. He had his headquarters at Smith Park, what is now Hawley Lake, and continued to build his sheep holdings.

Jaques would run both sheep and cattle once he married Beulah A. Whittemore, a cattle rancher, in 1913. The couple would graze their sheep and run cattle over seven townships. Their holdings would reach twenty-seven thousand sheep and five thousand cattle. Beulah worked side-by-side with Sant through both good and bad times. During World War I, the ranch had trouble getting men to work for them, and there was a shortage of materials. This problem was not unique to Jaques as other sheepmen were also dealing with short-ages of men and materials. Jaques traveled many miles by horseback throughout the White Mountains as his 1913 Ford could only go where there were roads, and at this time, there were few. Getting gas or even having a gas station to buy his gas could be a challenge.

In one incident, he mixed kerosene with gas to get him to where he needed to go. "Sant and Beulah man-aged their joint herds wisely, and for a long period their operation was the biggest wool shipper, the biggest taxpayer, and one of the biggest employers in Navajo County."[125] It is probably because he was the biggest taxpayer and employer in the county that he was able to get things done. Because the population was small, electrical companies did not want the expense of bringing electricity to the mountains. In 1936, the White Mountain Electric Cooperative was formed to do just that. With the outbreak of World War II, the cooperative failed because there were no materials to begin the process of building the necessary lines to bring in electricity, but, with a handful of determined citizens, one of which was Jaques, the Navopache Electric Cooperative was formed in 1946. Sant would be the first president of the cooperative, and it still operates today, covering ten thousand square miles. He retired from the sheep business in 1950 and went on to pursue other activities in the Phoenix area, but he returned to his beloved White Mountains in 1955, building a home that overlooked the Show Low Valley. "Sant Jaques has seen the White Mountain country progress in his eighty-three years of life from uncharted wilderness to the progressive vacationland, forestry products center, ranching hub, and residential area that it is today."[126] Jaques died in 1967 at the age of eighty-five. Sheepmen were concerned with more than only their industry in this state, as Jaques proved.

---

125 Ibid., 80
126 Pollock, 80.

The Dobsons

One family that came to Arizona in the late 1800s did not start out as sheepherders, but by the late 1900s had reportedly one of the largest sheep herds in the state. The first Dobson to arrive in Arizona was Wilson Wesley Dobson, who came from Ontario, Canada, in 1886. On a visit back to Canada, he told his younger brother, John, that the only thing lacking for growing agricultural crops was water. John joined his brother and worked for one dollar a day plus board. He was able to save enough money to buy his own land, but it wasn't long before he became discouraged when the Salt River would flood, destroying his crop. He returned to Canada in 1900.

While he was back in Rouleau, Saskatchewan, Canada, he married his childhood sweetheart, Sara Electa Scott, and began a mercantile store, but word reached him that the Roosevelt Dam was going to be built on the Salt River to control the flooding that had discouraged him from continuing to farm earlier in the Salt River Valley. With his friend, Alex Knox, John returned to the Salt River Valley, and they purchased 320 acres together.

Over the next twenty years, John would begin one of the largest ranching operations in the Salt River Valley. By the 1920s, he was a banker and board member of the Salt River Valley Water Users Association. John was elected president of the association in 1930 and served a two-year term. He had large land holdings in Mesa, Tempe, and Chandler. He ran twenty thousand head of cattle and grew alfalfa, wheat, barley, and sorghum in the

Salt River Valley. He expanded his operation into sheep in 1929, buying the George Scott Sheep Outfit, which gave him ten thousand sheep. He leased more land, with grazing rights in the desert and White Mountains from when he had purchased the George Scott Sheep Outfit. Unfortunately, this purchase was made just prior to the Great Depression of 1929, but like so many other sheepmen during this time, John Dobson survived. His alfalfa fields allowed his ewes to graze and lamb during the winter months since the rains would not allow for harvesting the alfalfa hay. He would eventually split the sheep outfit between his sons, Roy, Cliff, and Earl, plus his son-in-law, Larkin Fitch.

John and his wife, Sarah Lecty, had five children. During World War II, John split his ranches among his five children, Roy, Cliff, Earl, Mildred, and Ruby. The two oldest boys received the majority of the acres. Each had their own sheep operations: Roy from 1937 to 1963 and Cliff from 1937 to 1978. Earl began to herd sheep in 1937 also. John's son, Earl, had three sons who were all involved in the sheep business at one time. His three sons were Dwayne, Dennis, and Vinson. Dwayne, their eldest son, was the last of the family to give up raising sheep. He sold his sheep and all his grazing rights in 2011 to Joe Auza Sheep Company, keeping a hundred of the sheep just for the family. When the author visited with Mr. Dobson, he only had about ten or twelve sheep left. He was in the process of divesting himself of all the sheep.

In an interview[127] with Dwayne and Vinson in 2006, the brothers talked about their family's involvement in the sheep business, beginning with their grandfather, John. In that interview, Dwayne stated that the ewes lambed in late October and into November on alfalfa fields in the southeast valley, in Chandler. The ewes and lambs wintered in the desert area until April. Then the lambs would be sold off the first week of April and go to Colorado mostly, but some went to feedlots in Texas. There, the sheep are fed a high-concentration feed that is mostly corn so they will gain forty to fifty pounds before they are slaughtered. Most years, Dwayne said that he could sell the lambs, but the price was down in 2006, and he had to place the lambs in a feedlot in Colorado. The ewes were then trailed north, which can take upwards of fifty days, to the ranch. Most years, they arrived in the White Mountains, in Greer, about June 1 and stayed until mid-August. Dwayne said that they sheared the sheep prior to herding them back down the Heber-Reno Trail in late August. Shearing twice a year started in the 1980s because the Dobsons felt that the sheep were easier to handle during lambing.

Continuing the oral history given for the Arizona centennial, Dwayne remembered the times he helped his father during the lambing season by taking care of what he called the "leppie" lambs or those that were orphaned. During a conversation with the author, he also spoke of

---

127   Vinson and Dwayne Dobson interview, *Chandler,* September 12, 2006, www.chandleraz.gov /content/Oral_history_collection_list.pdf

these times fondly. Sometimes the ewes would die, and the lambs would have to be hand-raised—meaning bottle-fed—or the ewe would reject one of the twins and that one would need to be bottle-fed. Bottle feeding was at least four times a day and sometimes five. If a ewe lost her lamb, she would be full of milk and would be paired with an orphan lamb, but she would need watching to make sure the lamb was getting enough milk to grow, get healthy, and adopted by the ewe. If twin lambs were born, the two lambs would be hobbled together so that the weaker one would learn to keep up with the stronger one. The lambs would grow with their mothers on alfalfa pastures or be moved to the desert pastures if the rain had been good that summer and produced vegetation that the sheep could feed upon until sold in April as spring lambs.

Dwayne also remembered many chores that he had to help with, besides taking care of the newborn lambs. Much needed to be done, especially before mechanization. With cows needing milking and other farm chores, he was a busy youngster. Vinson told similar stories, but being fourteen years younger, he remembered more mechanization on the ranch.

The Dobsons were also interviewed for an *Arizona Highway*[128] article. One of the foreman, Byde Hancock, spoke of them trailing sixty-five thousand head of sheep in the 1930s. He said that a ranger from the Forest Service would

---

128 Joan Balza, "Last of the Old Sheep-Drive Routes: the Heber-Reno Trail," *Arizona Highways* (November 1986): 14–20.

accompany them to ensure that the sheep and herders stayed on the Heber-Reno Trail and stopped at designated camps. Hancock worked as a foreman for fourteen years, and in the article stated that he had seen many changes in the sheep industry. He believed that trailing sheep was about to end, as did other sheepmen. With increasing costs and subdivisions and shopping centers replacing the agricultural fields once used to graze the sheep in the winter, many sheep outfits had gone out of the business Hancock continued. He also said that cheaper imports from Australia and New Zealand were taking a share of the market away from the American sheepmen.

In the same article, Dobson's children all talked of their first jobs working with the sheep. Earl, John's youngest son, told of his first trip on the Heber-Reno Trail when he was eighteen. His responsibility was to supply the sheep camps. Dwayne's son, Carey, was eighteen when he decided if he was going to take over the business someday, he had better go up the trail. He was the camp tender, the man responsible for setting up the camp. He was also responsible for packing the burros and cooking. With little understanding of Spanish, and not knowing the sheep trail, he relied on the two Mexican herders. The burros seemed to know where the camps were going to be set up each day on the trail.

From the oral history and *Arizona Highways* article and talking with the Dobson family, it was obvious that sheepherding was a passion for them, and they enjoyed the times that they spent out on the trail with the flocks.

Twins hobbled together. *Photograph courtesy of author.*

This was a common theme for all the families, whether Basque or other nationalities. Peterson wrote, "To the Basques, raising sheep is not just a business. It is much more than a financial venture. It's a way of life, but it's a way of life that's quickly disappearing." [129] Dwayne Dobson summoned it up this way: "I don't have to be in the sheep business to make a living anymore," he said. "It's more of a family tradition than anything, lasting eighty-two years."[130]

The Pouquettes: Joseph, Leon, and Albert[131]

Pierre Pouquette, father of Joseph, moved his family to the United States from France in the late 1890s. They first settled in California, and in 1888 Joseph was born. Pierre was in the sheep business near Ventura, California. When Joseph was two years old, the family returned to France. They returned to the United States when Joseph was about seventeen. He settled in Ash Fork. In 1915, Joseph went to California and married his childhood sweetheart, Marguerite Bordenave. Marguerite's family had come about the same time to California, and they settled in San Francisco. Joseph and his bride return to Ash Fork and Joseph purchased three thousand sheep.

Joseph; his uncle, Simon; and a hired man named Bargo traveled to Panguitch, Utah, to purchase four thousand yearling ewes in 1917. They trailed them down to Lee's Ferry at the Colorado River. Because the ferry could only

---

129 Sue Peterson, "Shepherd of the Open Range," *Arizona Highway* 54 (1978):2.
130 Personal communication with Dwayne Dobson, August 23, 2016.
131 Personal communication with Edith Pouquette, October 2016.

hold a hundred sheep each trip, it took forty crossings to get all the sheep across the river. This is one of the latest dates for sheep to be trailed from Utah into Arizona. The majority of the sheep came during 1902 to 1908, with 1905 seeing fifty thousand sheep and 1906, eighty thousand. These yearling wethers were fattened, sheared, and sent to market. There is no record of sheep coming from Utah between 1908 and 1917, when Pouquette brought the four thousand into Arizona. Many sheepmen did receive rams from Utah later, but these were not trailed. Albert Sr. and Joe Manterola both bought Suffolk's rams from Utah.

Joseph was in the army the last two years of World War I. He was stationed in France as an interpreter because he spoke French. After being discharged from the army, Joe moved his family and his sheep business to Williams in 1920. He had summer grazing range nearby, and in the winter, he and his sheep could be found in the Wickenburg-Congress area. He told someone within the AWGA in 1967 that "we used to go on the train—moved kids, chickens, wood stove, and everything—while the sheep grazed back and forth on the trails." This occurred twice a year as the sheep were moved between summer and winter pastures. He further recalled that in later years, the family moved by car between Williams and Wickenburg, and it took two days as they would stay over in Prescott.

Joseph remembered good and bad years in the sheep business. Sometimes lightning would kill some of the sheep. His worst time came in 1949 when there was an early snowstorm in the northern portion of the state. The crews in the city of Williams helped him break a trail for the sheep so they could move from the Red Hill Ranch to Ash Fork. It took twenty-one days for the animals to move thirty miles!

Typical herder's camp near Ash Fork circa 1937.
Photograph courtesy of Edith Pouquette.

In 1967, Mr. Pouquette was considered the oldest living sheep owner still in business in Arizona. In his fifty-one years in the business, he saw many changes. One he remembered was the change in food demanded by the herders, those who tended the sheep daily. In the early days, the herders were given rice, raisins, and beans as their main staples. When the story was written for the AWGA, he shook his head and said that the sheepherders now demanded canned ham, cabbage, green salads, and drinking water hauled from town.

A year before his death in 1968, Joseph was still running six thousand head of sheep and two hundred and fifty head of cattle with the help of his two sons, Pierre and Albert. He was asked if he preferred sheep to cattle and his response was, "Well, there is more work to the sheep, and that's what I like."[132]

Albert Pouquette Sr.

In personal interviews via telephone and in person, the life of Albert Pouquette Sr. as a sheepman was told. Albert was the youngest son of Joe and Marguerite. His brother, Pierre, was also a sheepman. Marie was the middle child. Albert and Pierre are third-generation sheepmen. There would be one more generation involved in the sheep business, and that would be Albert's son, Albert, Jr. Albert Sr. worked with his dad in the sheep business starting in about 1946. The year before that, his father, Joe, and his sister, Charlotte, purchased what would be called the Red Hill Ranch. It had been owned by Joe's uncle with a man by the name of Manuet. They sold it with eight thousand sheep to Charley Burton, who was able to pay off the $100,000 note in two years. Burton owned and operated the ranch until 1928, when he defaulted on another loan,

---

132 Arizona Wool Growers Association, NAU.MS.233, Cline Library. Special Collections and Archives Dept. Series 1.4.15. Part of two pages of a report that may have been part of the annual report for the association. The two pages are marked three and four with no date. However, it would have to be prior to Joseph's death, as the report stated he was still running sheep with the help of his two sons, Pierre and Albert.

and everything was sold. It changed hands many times between 1928 and 1945. The Pouquettes named the ranch the Red Hill Ranch, and the sheep company obviously was named after it. This is where Albert Sr. comes in. He helped his dad and older brother run the Red Hill Sheep Company. Albert married Edith in 1947 and they had four children. His son, Albert, worked with him running the outfit. Albert and Edith had a winter home in Peoria. They grazed sheep on an area west of Peoria called Marinette. This continued until about 1960 when the land was sold to Del E. Webb for $10 million. Many sheepmen who grazed this land had to find other winter pastures. The Atchison, Topeka and Santa Fe Railway ran a line along here. This made it easy for those no longer trailing their sheep from summer pastures in the northern portion of the state to have their sheep railed into the valley.

The railroad was used until the 1970s, when trucks were better suited to take the sheep from winter grazing land to summer grazing land in the same day. At this time, the railroad declined the shipment of animals. While this is not just germane to the Pouquettes, the railroad had its problems in shipping animals. In numerous correspondence found in the various sheep outfit folders at the AWGA, there were many complaints of injuries or death to the sheep while they traveled by rail.

Albert and Edith moved their winter operation to Blythe, California. The older children, Albert Jr. and Jerry, would attend school here. The boys spent many days with the sheep and sheepherders. Albert Jr. told the author that he learned the business from the ground up. He told of he and his brother following behind the sheep and throwing dirt clots and how the sheepherders would be mad. The boys, their mother recalled, had to be about six or seven. Because the boys were small, their father said they looked like coyotes to the sheep, and thus the sheep would scatter. Edith said that sheepmen and their families like lamb meat, and there would be times they would kill one of the sheep. She continued her story, saying that she would sit on the sheep while Albert Sr. would slit its throat. The animal would be hung from their rafters after being gutted for a day or so and then, just like those who hunt deer, the animal would be cut up and the meat frozen. Edith wrote, "The cooks were the best and of course lamb was always on the menu."[133] The family moved to Ehrenberg and built a house when the two boys would be ready to attend college in a few years. The family moved to Williams full time when the youngest daughter, Chedelle, started high school.

Albert Sr. sold the Red Hill Sheep Company around 1984 to Joe Auza of Casa Grande. This ended the fourth generation of Pouquettes as sheepmen. Albert Jr. told the author, "It was a good life." Albert Jr. worked with his father and his uncle Pierre until the sheep were sold. They continued running cattle until Albert Sr. and Uncle Pierre both died. Albert Jr. continued in the cattle business until 2000, when he too sold out.

---

133 Personal communication with Edith Pouquette.

The AGWA has a little information about another Pouquette, Joseph's brother Leon. Mrs. Edith Pouquette wrote the author that Leon and Joseph married two sisters. Leon married Amilie. Leon and Amilie had two children, Yvonne and Felix. Yvonne married Pete Espil, another sheep rancher. Leon ran an outfit with John Aleman called the Aleman Pouquette Sheep Company. When Leon died in 1942, John Aleman ran the five thousand sheep outfit with Felix helping during school holidays and summer vacations. A letter to the Wool Growers Association written November 19, 1945, requested that Felix be discharged from active service as he was needed to take over the sheep business since John was in bad health. It is unknown whether Felix was discharged. Edith Pouquette remembered when she and Felix played as children, but doesn't remember him working for Mr. Aleman.

The Basque

Many nationalities participated in the sheep industry, but none more so than the Basque, who arrived both from the area of the Spanish and French Pyrenees. While Basque herders were probably in the state from the 1860s to the 1890s, most were probably herders for other sheepmen. During the 1890s, Basque names begin to appear in the literature, and many of the Basque families in the state today can trace their families arrival beginning in the late 1880s. The Basque (from an adjacent area of Spain and France) have been part of the Phoenix metropolitan landscape for the past hundred years and further back in time in the state. "Not all woolgrowers are Basque, but Western

Pouquette's donkeys loaded before heading out on the trail.
Tents were draped over the boxes and then other protective
shielding added. *Photograph courtesy of Edith Pouquette.*

woolgrowers could not operate apart from the skill and
sheep savvy of Spanish and Basque shepherds."[134] And, in
fact it has been recognized by some historians that *Arizona*
is a Basque word. It is widely believed that *Arizona* is de-
rived from two Basque words meaning the "good oaks."[135]

The Basque came to the Americas, the United States,
and Arizona for many reasons. Up until the civil wars in
the 1800s, the Basque were isolated from the politics of
either France or Spain. With the wars, the Basque began
to be inducted into the military. Rather than serve, many

---

134 Ruth Nourse, "First Lady of Sheep-Growing Clan Thinks and Cooks in Two
Languages," *The Arizona Farmer-Ranchman* (October 19, 1968), 36.
135 Jim Turner, "How Arizona did NOT Get its Name," Arizona Historical Society.
Archived from the original on August 1, 2007. Retrieved May 15, 2016. Donald Garate,
"Arizona, a twentieth-century myth," Journal of Arizona History 46(2), (2005): 161–184

of them emigrated or deserted. Some Basque left after serving. Lack of tillable land, a drought in the first part of the 1900s, overpopulation, and poverty are all functions of a small land area, and as families grew, the eldest son was the only one to inherit the land, forcing younger brothers to migrate. They first immigrated to South America, taking a variety of jobs not always associated with sheep. With news of California's gold rush, many decided to make their fortunes in the American West. Many from Argentina left for California by the way of ships departing from the west coast of South America. Those who left Europe for the United States had to sail by ship around the tip of South America to reach California. Unfortunately, they were discriminated against in the mines, and many found it easier to become shepherds who sold mutton to the miners.[136]

Two other reasons brought different ethnic groups to the United States: the Homestead Act of 1862 and the Desert Land Act of 1877. These offered immigrants the prospects of land and prosperity. The Homestead Act of 1862 encouraged individuals, families, and immigrants who intended to be naturalized to move westward and claim 160 acres of public domain. Within five years, improvements had to have been made, and they had to live on the land. If both criteria were met, title was given to the land. The Desert Land Act allowed 360 acres to be purchased at twenty-five cents an acre. It was felt that given the arid conditions in

---

136 William Douglas and Jon Bilbao, *Amerikanuak: Basque in the New World* (Reno, Nevada: University of Nevada Press, 1976), 201–212.

the West, the 160 acres of the Homestead Act was too small for an individual or family to make a living. The land had to be irrigated within three years. Many of the sheepmen that were researched used one or both of these to acquire land in the United States.

Before the continental railroad was finished in 1869, the Basque arrived in the American West by way of South America. Preliminary research showed more information has been found concerning those who arrived in the United States via New York. Immigrants faced many challenges, such as crossing the ocean as travel time depended on weather conditions, the process of immigration—health checks being one of the biggest—and a new language. While many Basque spoke their own language and either French or Spanish, or both, English was foreign to them. Some Basque never made it to the West as they were tired of traveling by the time that they arrived in New York. Today, a large population of Basque calls the Brooklyn area home.

When a Basque arrived, he sometimes was met by an employee from the Santa Lucia Hotel shouting in Basque, "Are there any Basque?"[137] Those Basque would then be taken to the hotel and would receive help in making their preparations to move westward. Those who spoke no English or only a word or two would have instructions pinned on their clothes for conductors to help them to move along the rail line from train to train to get to their final destinations. Some of the Basque families interviewed

137 Ibid.

stated that their forefathers only knew a word or two of English, usually food related. They could say "hamburger" or "ham and eggs," and that would be what they ate for the entire trip across the country! One family stated that their father never ate hamburger again! One Basque was unable to eat the whole trip across as he did not have a tie, a requirement to eat in the dining car on the train. He could not make himself understood so he went without. The men came one or two at a time with empty pockets and hope within their hearts, seeking a new beginning in new land far from all they knew!

Jean Pierre "Pete" Espil

One of the earliest Basque to arrive in Arizona after spending seven years in California was Jean Pierre "Pete" Espil. Many Basque would work for Pete during the years and get their start in the sheep business because of him. Like other Basque, Pete left his home in France with his cousin, Martin, as the prospects were not good for either one of them. Pete had been born in Bagneres-de-Bigorre, Hautes Pyrenees, France, in 1870. Neither one was the oldest, so they would not inherit the family farmstead. The family told the author that their grandfather, Pete, never knew his correct age. He may have been eight, ten, or twelve when he and Martin stowed away on a ship heading to New Orleans, a logical place to arrive in the United States since there would be other French speakers. (Another story says the two worked their way across the ocean on a cargo ship.) They traveled to Los Angeles by train and then took a stagecoach to Sacramento. The granddaughters said that once they landed in Los Angeles, he only had a few English words to get by with: "fried eggs and ham." Once his English improved, he never ate fried eggs and ham again.

Martin had a job lined up before he had left France and went to work in the sheep business. The weather was too wet for Pete so he headed south to the Long Beach area. His first job was with the sheepherding outfit Miller and Lux Land and Livestock Company. This gave him good experience but only a twenty-dollar gold piece when he was laid off, even after seven years of working for the outfit. Mr. Lux had a problem with gambling and lost all the sheepherders' pay in a card game. Martin remained in California, but Pete boarded a train heading east, and the family has wondered if he was thinking of returning home.

Upon his arrival in Flagstaff, the environment with the mountains surrounding the town made him feel like he was back in France. When Pete came to Flagstaff is uncertain as two different dates are given in the literature, so for ease the 1890s will be used. The actual day and year is not as important as what he contributed to the sheep industry in the state. When he first arrived in Flagstaff, he met Harry Embach, and the friendship between the two men remained for the rest of their lives. Harry had recognized him as a Basque sheepherder and helped him secure a job. Embach was the accountant at the Babbitt ranch. It is unknown how long he worked at the Babbitt ranch, but sometime later, Pete went to work for Hugh Campbell. Like in California, he trusted his employer, who was also a banker, to keep his pay until he requested payment. Unfortunately, the banker absconded with $80,000 of the bank's money and Pete's pay. The bank asked Pete if he would run Campbell's sheep until they could be sold. The ewes were used to offset part of Campbell's debt, and Pete received the lamb crop for his services. The Espil Sheep Company was born.

Pete, with three witnesses standing beside him, became a citizen in 1899 at the Superior Court of San Francisco. In 1902 he was able to secure summer range of 175,000 acres for his sheep with the US Forest Service. This was a ninety-nine-year lease. A cabin was built at Reese Tank and became his headquarters. Like earlier sheep and cattle ranchers, he found that water was necessary for his sheep's survival, so with the help of Frank Auza Sr., he built a metal tank at Pat Springs and a water pipeline in 1926. This pipeline was built by hand. It ran from Pat Springs to the ranchland. He added two other water tanks closer to where he grazed his sheep as his herd grew. Pete brought a section of land from the Otondo family near Lockett Meadow in the foothills of the San Francisco Peaks. A dipping vat[138] built by Pete can still be seen in Schultz Pass. Espil added the Deadman Ranger Station to his holdings and converted the station into a home for his growing family. In 1913 or 1914,

---

138 For an explanation of the use of a dipping vat, please see the conclusion to chapter two.

Pete Espil, JP "Pete" Espil, Louis Espil, and a wool buyer.
Photograph courtesy of Elizabeth Espil Mooney.

Pete married a Spanish-born Basque girl, Isidora Aristoy, who had come to the United States in 1913 when she was twenty-one. They had two sons, M. P. (Pete) Jr. and Louis, and one daughter, Dora, who was the middle child. He partnered with M. I. Powers for a few years until Powers sold his interest to Ysi Otondo in 1927.

Early in his career, Espil was instrumental, along with many other sheep ranchers, in purchasing a mile-wide swath of land along what would become I-17 between Phoenix and Flagstaff. That land was used to trail sheep between the fall and winter grazing pastures each year. This swath of land kept the sheep from being trailed on private land and crossing fence lines. When the freeway was built, bridges allowed the sheep to pass under the highway.

Pete's citizenship papers were destroyed in the fire that resulted after the San Francisco earthquake of 1906. While he was visiting with a friend at a Flagstaff saloon, the foreman of another ranch overheard his concern. This rancher used the federal authorities against Pete, and he would lose his land because they could then acquire this much-desired ranch land. Three years and several thousand dollars later to attorneys and private investigators, two of the three citizenship witnesses were located in Aurora, Oregon. Only a portion of the Espil Ranch was lost. When it was all over, the granddaughters told the author, he said they all had to work together, and he never made an issue of the loss he suffered.

Later in life, Epsil became one of the largest sheep owners within the state, basing his operations in Litchfield. Here he leased winter grazing land from Goodyear Farms. He had eight thousand Rambouillet ewes. Pete ran a double operation by having feeder lambs, thus more than doubling the number of sheep he had. His sons, Louis and Pete, later joined him in the business. He used various rams in his operation to ensure the maximum number of lambs born each year.[139] He purchased rams known to be meatier from Mr. Burton of Idaho, making his sheep heavier. His lambs were sold as milk-fed spring lambs with the top weight being between 120 to 130 pounds. The lambs

---

139 "Lambing and Feeding Operation," *Arizona Farmer-Ranchman* (November 23, 1957): 16.

were fed on their mother's milk and grazed on the alfalfa and barley pastures of the Goodyear Farms. His son Louie remarked that their profits have gone down over the years, and they were lucky to have a 6 to 8 percent profit. In the past profits were 20 to 25 percent. Many factors contribute to the percent of profit a sheepman can make. One factor is labor costs.

In a short biography written by the AWGA[140] on July 12, 1960 for their annual meeting, facts about Espil appear, differing somewhat from the information presented above. The main difference reported is Espil arriving in the United States in 1891, which would put his age at twenty-one. It was common practice within the AWGA to honor members who had passed since their last annual meeting with a short bio. Pete Espil had been the oldest active member of the AWGA until his death as he had joined the association in 1906.

M. P. "Pete" Espil

Michel Pierre Espil was born in Flagstaff in 1918. He grew up in the sheep industry and spent his school years following the seasonal sheep migration, along with his younger brother and sister, Louis and Dora. He would begin the school year at a parochial school in Flagstaff only to finish the rest of the school year from November to May at a public school in Wickenburg. From eighth grade to his senior year, he went to school in the Wickenburg area, where his family shared a home with his

---

140   Arizona Wool Growers Association, NAU.MS.233 Series 1.2.8 and Series 1.2.9

uncle Michel Ohaco's family. He excelled in basketball, baseball, and track at Wickenburg High School. In 1936, his senior year, he transferred to Glendale Union High School, as the Espil Sheep Company was now based for the winters in Glendale. At Glendale High School, he continued to play football and added an Arizona High School Football Championship. Pete Jr. went on to Arizona State, studied agriculture, and continued to play football and track. His father was injured during his sophomore year, and he returned home to take over the family sheep business. He attempted many times to return to college but never found the opportunity as there was always pressing business he had to attend to. When World War II began, the war effort needed wool and a meat supply. The military felt his place as a wool grower, producing the needed wool and meat was most important, and denied his attempt to enlist in the army.

In 1944, Pete married his high school sweetheart, Yvonne Pouquette, who also grew up in the sheep industry in Wickenburg and Williams, Arizona. The couple had three children, Michel Jr., Yvette, and Denise. During the 1950s, 1960s, and 1970s, the Espils would make their sheep business one of the most efficient in Arizona if not in the West.[141] They added more land to their father's original land holdings during these years, and in 1977, they converted the ranch to cattle. The change was necessary due

---

141 Paul W. Pollock, "M. P. (Pete) Espil," *American Biographical Encyclopedia Volume 2* (Phoenix, Arizona: Paul W. Pollock, 1969), 120. This section was written with the help of Pete's son, Michel Jr.

to changes in government regulations that prevented them from defending their flocks from predators and from a lack of qualified labor to tend the sheep. A third reason for switching to cattle was the importation of Australian and New Zealand meat and wool. The Espil Ranch continued to run feeder lambs on alfalfa pastures in the Phoenix area during the winter, in addition to the steer operation in Flagstaff in the summer. In August 1986, the Espil Ranch was sold to the Navajo Nation, allowing Pete to enjoy his retirement while spending winters in Litchfield Park and summers in Flagstaff, Arizona.

Pete was elected in 1944 to the board of directors of the AWGA and continued in that capacity until he was elected president in 1958, serving until 1969. In 1969, he was elected vice-president of the National Wool Growers Association, and during this time, Pete also was on the board of directors of the Western Range Association.

Pierre, one of Dora's two sons, remembers stories told about his mother. Dora was Pete Sr.'s only daughter, and like many of the fathers during this time, he was very protective of her. She was not allowed to take risks like her brothers were, and she wasn't even allowed to ride one of the horses. Dora's answer to the problem was to sneak out to the barn and take a "thrilling horse ride."

Dora spoke of the Depression and the financial stress on the family. She mentioned that while in the town of Flagstaff, she passed a bank that had a black wreath on the front door with had a sign that stated, "Gone but not forgotten." The Depression made a lasting impression on young Dora's life. Dora spoke of moving down from the high country during the winter months to their family home in Wickenburg. During the winter months, she went to school at Garcia's Little Red Schoolhouse. Their Wickenburg home faced the road to California, and she remembered the large number of people traveling because of the Dust Bowl. The life of a sheepherder's family was demanding, yet their family had it far easier than most mainly because of what they did for a living. They always had food on their table while many others went without. Those who stopped and asked for food were always welcomed and given a meal before they continued to journey westward. Dora did manage to convince her overprotective father to let her attend the Colorado Women's College in Denver, Colorado. A year later, she transferred to the University of Arizona, much closer to home, and completed a degree in business administration. Following this event, she met Herbert Prouty on a Valentine's Day blind date. They married, moved to Colorado, and raised two sons.

Louis Albert Espil

Louis was the youngest son of Pierre and Isadore, born May 27, 1922. Like his brother before him, he was a great athlete, excelling in football, basketball, and track and receiving many accolades. He enrolled in agricultural classes at the University of Arizona and became a member of the ROTC. A physical disability prohibited him from joining the military when World War II broke out. Between 1942 and 1946, Louis worked for his uncle, Mike Ohaco, on his sheep ranch. The four years of experience gave Louis much insight into the sheep business, which he and his brother put to use in growing the Espil Sheep Company. Initially the company had about three thousand, two hundred sheep but it soon grew to about eight thousand by 1974. Louis would ride the vast holdings of the summer range that was most of the San Francisco Peaks minus the south face where Flagstaff is located.[142] Sheep require more attention than cattle so something always needed tending.

Pablo Uruttia, a sheepherder who worked for Louis Espil. Pablo saved the lives of the Louis Espil's four daughters and the two Arambel boys by killing a mother black bear when they had unknowingly walked between the mother bear and her cub. *Photograph courtesy of Liz Espil Mooney.*

---

142 Paul Pollock, "Louis Albert Espil," *American Biographical Encyclopedia Volume 3* (Pollock: Phoenix, 1974), 24. Personal communication with Liz Espil Mooney, October 10, 2016.

Louis was very active in the industry. He served for eighteen years on the Arizona Livestock Credit Association, which provided monetary assistance to farmers. He was a director for the AWGA and was active in the national organization. He was an advocate against the Wool Act of 1954, believing "government subsidies... undermines the independence of the ranchers and encourages government intervention."[143] The Kiwanis of Arizona awarded him the Agriculturist of the Month in 1954 for his dedication to agriculture and his community service. Because sheepmen need to use the national forest to graze their sheep, he was on the Advisory Board for the Coconino National Forest as the representative for the range people and voiced their concerns about how public lands were administered.

Louis married Marion Ansley in 1950, and the couple had four daughters, Liz, Luanne, Laurie, and Margaret. One story that his daughters sent to the author explained how they loved to see the sheep on the mountains around the San Francisco Peaks. The girls may not have accompanied the men and sheep on the trail, but they understood the hazards of trailing. They were concerned for the "sheep's safety, so seeing the sheep on the mountain after the long walk from the valley was a beautiful sight indeed."[144] The girls would, whenever possible, visit the herder and camp tender, taking them supplies and eating their delicious food. "It was an

---

143 Ibid., 24.
144 Personal correspondence with Laurie Espil Goode, October 13, 2016.

unforgettable memory, looking out all the ewes grazing on the Peaks and listening to them."[145]

The Echeverrias: Miguel, Matias, and Fermin[146]

The three Echeverria brothers, all from Spanish Basque, arrived at different times into the United States. The brothers would marry Basque girls, and their extended families would intermarry. This would begin a long period of Echeverrias in the sheep business of Arizona.

Miguel, known as Mike, arrived first in 1903. He was the oldest of the three brothers. Miguel was born in the small village of Viscarret, Navarra, Spain. At a young age, he was in charge of the family's sheep. Being the firstborn son meant he would inherit everything, but he knew that his opportunities for fame, fortune, and success were on the other side of the ocean, in America. The family today cannot say how Miguel's parents were able to raise the money for his passage to America.

He first went to Riverside, California, and worked for John Pierre and Louis Meten. His salary was $15 per month. Just like as he did in his home country, he herded sheep high in the mountains of California, and his only contact with another human was with the man who delivered his camp supplies to him. While the name of this man is not known, Miguel told his family this man was responsible for him learning to read, write, and do math.

---

145 Personal correspondence with Luanne Espil Robertson, October 13, 2016.
146 The Echeverria section was written with information supplied to the author by Irene Aja and personal conversations between Irene and the author.

Miguel decided to head to Arizona when he learned that the sheepherder was making $35 per month.

He arrived in Winslow in 1907. He saved all his money, and when he had a $1,000, he pooled his money with Mike Ohaco. They asked for a loan from the Central Arizona Bank to purchase a water-right base near Ash Fork that included several hundred ewes. In a tribute to Miguel and his wife, his daughter Yvonne wrote a detailed "tale" about her parents for a family reunion. She wrote of her father, "Those were the days," he said to me (in Spanish, of course), "when bankers had faith in youth and didn't require so much collateral. I was lucky to make money before income tax came into being. Now the boys have to pay so many taxes." He was very successful and was able to prosper, buying and selling sheep and forming new partnerships with other herders.

In 1912, Miguel contracted a mysterious illness and returned to Spain to die. However, a Basque *curandero* gave him a cure, and he returned to America once again. He continued to raise sheep, buying and selling more. His life was about to change in 1919, when two twins, Angelita and Vicencia Erros, arrived from Espinal, Spain, a neighboring village to Miguel's.

The Erros had lived for generations in this village, just as the Echeverrias had in their village. A younger sister, Maria, was born, changing Vicencia's life. As a young girl in a normal family, she would have learned household chores, but her parents began to treat her like a boy, teaching her all the tasks normally assigned to a boy in the family and few household chores, like cooking. She attended school through the eighth grade when she was then hired to other families and performed household jobs, such as doing the laundry and cleaning the stables and dog kennels for very little pay. Her only fun was at fiesta time, and she was known for her dancing ability, the *jota* being her favorite dance.

The two oldest sisters made their way to America, and the family does not know how. The two girls arrived in Flagstaff and went to work for their mother's twin brother, Gregorio, who was partial owner in a hotel there.

Vicencia loved her evenings, when she could go into town with the other maids. With her limited English, she ordered ice cream. Yvonne wrote, "She was embarrassed, years later, to realize that her two new English words did not convey real meaning. I guess she ate vanilla for months as she happily responded, over and over again, 'ice cream' to the clerk's repeated question, 'Which flavor?' Finally, the exasperated clerks would give her a plain ol' vanilla cone. Her world broadened considerably when she learned the word *chocolate*!"

In late 1919, Miguel stayed at the hotel. He took one look at the two sisters and promptly asked Vicencia out on a date. He wasted no time in confessing his love for her and asked her to marry him on their second date. Love overcomes many things, and it did here too. When Vicencia was growing up in Espinal, Spain, she had vowed never to marry anyone from Viscarret as they were arrogant and rude! It would be six months before they were married as Vicencia had to obtain permission from her father, and this was done through a letter. Her uncle also gave his permission. In July 1920, the couple married in a grand style for that year. Miguel paid approximately $1,000 for the wedding, and their daughter Yvonne said her mother always said that since they only received a cut-glass bowl, Miguel had paid a great price for it!

With Vicencia barely seventeen, she quickly learned to cope with learning to cook, take care of other household chores, and tend to the arrival of their first child, Josephine, all while Miguel was tending to the sheep. Vicencia learned to cook with the help of her friend, Serapia Echenique.[147] Two more children would arrive shortly, one of them a son, Mike, and thus the inheritance issue was resolved. With the three children and wife, Miguel decided to return to Spain and live a life of leisure after all the hard years spent in the sheep business. Vicencia was happy to return to their native land as she had been missing her family. They would only spend one year in Spain and with four children return to the United States and Arizona. Miguel had grown restless playing cards and telling stories of his adventures in the States.

They settled in Wickenburg, Arizona, where other children would be born. Wickenburg would be home for them

---

147 There was a Jose M. Echenique who may have been her husband. He was born in 1887 on the Spanish side of the Pyrenees. He arrived in the United States in 1905. In papers at the AWGA, Mr. Echenique requested grazing permits in several counties with the earliest being 1938. He also paid to the AWGA in 1941 for his pro rata cost of using the Bear Springs, Mud Tanks and Beaverhead Driveways for running about 5,200 sheep. He had sold his sheep in October 1947 along with his Buck Springs Allotment the following fall, 1948.

for fifty-eight years. Miguel and John Aleman purchased the Cross Mountain Sheep Company, headquartered near Seligman, from Campbell and Francis. This gave them four thousand prime ewes and two thousand yearlings. With other costs, they paid $32,000. The Depression was only a few years away, and through poker winnings, they bought groceries and paid the debt, at least the interest, Yvonne wrote. She continued that she was in possession of his set of books. "It consists of one small 4x6 booklet issued by Valley Bank, in which Dad had recorded sheep counts, bank account balances, and a grocery list for the herders, all neatly tabulated, one account per page! Try to do business that way these days! He always kept this booklet in his shirt pocket, alongside the hundred-dollar bills he always carried," Yvonne wrote.

Miguel retired from the sheep business in 1958 when he was seventy-three years old. John Aleman sold his interest in the Cross Mountain Sheep Company to the oldest son, Mike. Three other children would also become part owners in the company, Bob, Rudy, and Julio. This allowed Miguel to retire. Mike would eventually move on to have his own sheep company.

In 1996, Yvonne wrote for the Echeverria family reunion how and why her parents came to the United States and what their life together was like. Through this document, much of the above information has been garnished. She told of the birth of her parents eleven children (there were twelve, with one dying two days after birth and how her mother only wanted four children, but dad wanted many flowers for his garden) and how the Basque culture was instilled in the family. Family reunions and birthdays were a big part of the family's life as they were times of camaraderie and great food. Vicencia told her children about all the relatives that she could. Her mother kept Basque culture instilled in the family. As Yvonne wrote, "It was she who vocally transmitted all those quaint Basque customs and traditions to us, diligently trying to superimpose them over the new American customs that her children were coming to accept. (Basque girls must marry Basque boys)." The family was served traditional Basque dishes such as *bacalao*, pig's feet in tomato and *morcilla*. Irene Aja told the author that *bacalao* is dried salt cod that is white, delicate, and tender once it has been rehydrated. *Morcilla* is also called blood sausage and is thicker than most sausage. It is stuffed with pig's blood, rice, onions, and spices. The children also learned from their mother that there was no such thing as idleness, something evident in the long hours her mom would be awake taking care of the main tasks for such a large family. This work ethic had been instilled in her when she lived in the Pyranees, and she would instill it in her children who lived in Arizona. Rarely was a recipe cooked the same way twice as she would experiment with ingredients that were on hand. She made many knitted and crocheted gifts for the children for birthdays and Christmas. Their mother would be the doctor, treating scrapes and illnesses. She attended the school functions. She taught the girls handy skills, but most of all, she taught the children to respect her and others, the value of truthfulness, an appreciation for their adopted home of

America, to love each other, to stand up for one another, and to have confidence and faith in themselves. We will see this in other Echeverria families.

While the mother spent most of the time with her children since the father would be found with the sheep, the father did instill traits in his children also. Yvonne wrote,

> It was Dad who taught us to "read people," to look before we leap, to think up viable strategies, to implement our plans, to do our very best, to carry our name proudly, to set realistic goals and carry them through to fruition, to seek zest and amusement in life, and to do honest deeds. It was Dad who encouraged us to laugh, to be playful, to outwit each other slyly. It was he who daily checked our mental agility, our physical agility, and our stick-to-itive-ness. It was he who held us to our word. It was Dad who expected us to think for ourselves. It was Dad who spoiled us with ice cream and candy, and it was he who also froze us in our tracks with "the look." It was he who let us count all his money (including the one hundred dollars), and it was he who confiscated any stray money lying around the house. It was Dad who exemplified kindness and integrity, and it was he who showed us what good husbands do and what good fathers do. It was he who smiled proudly every time he cast his eyes upon his children, no matter his age or their age. It was he who was self-taught and self-content. It was he who laughed with his eyes, finding amusement in most situations. It was Dad who delighted in a clever turn of a word or phrase. He taught us casino and poker and how to add. He taught us how to face life.[148]

She wrote about the little jokes her father would tell, and when Mom made a decision, Dad would back her 100 percent.

---

148 Written by Yvonne Echeverria for the Echeverria family reunion, August 16, 1996.

Matias, the middle son of Felipe and Josefa, left his home and journeyed to the United States. How the family paid for his passage is, like Miguel's passage, a mystery. However, the passage was paid in 1906, and he left his home and traveled by ship to New York. He then made his way by rail to Phoenix, Arizona. The next day he went to work herding sheep for the Campbell and Francis Sheep Outfit in the Wickenburg area for $30 a month. Before long, he delivered the supplies to the sheep camps. He continued in this capacity for several years, saving his money. Like his brother, Miguel, he pooled the money and went into business with Pierre and Mike Ohaco. They grazed their sheep out of Flagstaff near Lake Mary.

Wedding bells were in the air for Matias and Pierre Ohaco in July 1917, when they married two sisters, Marie and Gracianne Arango, respectively. It was a double wedding ceremony in Flagstaff. Just prior to his marriage, Matias had been drafted into the army to fight in World War I. Matias sold his shares in the business to his younger brother, Fermin. Pierre sold out at this time to Fermin's brother, Mike. After a honeymoon in San Francisco and Los Angeles, Matias reported to Fort Riley in Kansas by August, and his wife joined him. Matias served in the field artillery as a private. He was discharged in June 1918. The couple moved to Los Angeles for a year after his discharge. He never ventured into the sheep business again and never returned to live in Arizona. He continued in a variety of enterprises in California with Pierre Ohaco. He was a walnut grower, had a vineyard, and was an avocado grower. Later, he went into a partnership with his son in having an avocado nursery and custom tractor and orchard care and planting business. Why Matias never returned to the sheep business is unknown.

Fermin Echeverria. *Photograph courtesy of Irene Echeverria Aja.*

One more son would emigrate to the United States. The youngest son, Fermin, left by ship from Le Havre, France, arriving in 1910. He was sixteen. As with his older brothers, he had helped his parents by tending their flock of sheep. He helped harvest the grain and hay and helped in the small slaughterhouse his parents owned. His job was to take the meat to neighboring villages via burro. His passage was in steerage, and he had taken the food he would eat on the voyage. Fermin told his family he remembered seeing the Statue of Liberty as the ship he was on entered New York harbor. He often expressed his feelings about seeing this statue.

Fermin was processed through Ellis Island, where he had to pass a physical exam to enter the country. Once he cleared all processing,[149] he boarded a train and headed to Arizona. Fermin had sent a letter to Miguel that he was on his way, but Miguel never received the letter on time, and there was no one to meet him. Fermin walked around the train station in Phoenix, looking for his brother, until he heard some men speaking Spanish. Inquiring if they knew his brother Miguel, he was informed that his brother was far away and would not be able to find him. The man, Pedro Arrese, also Basque, convinced Fermin to work for him. Miguel had received the letter late and had sent Mike Ohaco to pick up his brother, but by this time, Fermin was long gone. It would be months before Fermin was reunited with his brothers, Miguel and Matias.

Sitting on a buckboard pulled by two mules, Fermin and Pedro made their way up into the mountains of Arizona. Fermin was given two burros, a canvas tent, a bedroll, a Dutch oven, an ax, and flour, sugar, coffee, salt, salt pork,

---

149 In the early 1900s, immigrants entering the United States had to pass a physical exam. Even a small defect like a deformed finger could mean a trip back to the country of origin. This happened to one Basque girl.

rice, prunes, raisins, and beans. His salary was $35 a month. These supplies were loaded on the burros as he would move the sheep when necessary to better grazing areas. He tended the sheep of Arrese's for two years and saved his money, as many of the Basque did. With $800, he invested in his brother's and Mike Ohaco's outfit. The money was needed to pay for the supplies their outfit would need during the winter months. Prior to this, Fermin's boss, Pedro, had given him a $5 a month raise, but Pedro reneged on the raise when he found out that Fermin had invested his money in his brother's outfit. For a time, he worked for Peter Espil but still kept his small interested in his brother's outfit. In 1917, Fermin purchased his brother's Matias business interest when Matias was drafted for the war. Mike Ohaco purchased his brother Pierre's interest, and Fermin and Mike went into business together for the next twenty-four years.

Other changes were in the works for Fermin. In 1918, Fermin met an American-born Basque girl, Benancia Erro Miranda, and after a very short courtship, they were married that same year. Fermin had remarked that he fell in love with Benancia at first sight. Benancia had come from California to help her uncle, Gregorio Erro at one of the two Basque boarding houses found in Flagstaff. While working for her uncle, Benancia would help other sheepmen by interpreting for business transactions. Benancia's mother had been born a short distance from Fermin's native village in Spain and had immigrated to California in the early part of 1890.

After their marriage in Los Angeles, they returned to Arizona. In the early part of the 1900s, an American who married a foreigner, a noncitizen, lost American citizenship. Under the Alien Land Act of 1921, an Arizona statute, only citizens or aliens eligible for citizenship could "acquire, possess, enjoy, transmit, and inherit"[150] property. How she could have lost her citizenship when she was born here seemed very unfair, but she did. She was now considered a subject of the King of Spain. Just like any other immigrant, Benancia had to complete all the requirements and regained her citizenship in 1926, but this was not the only problem of being a noncitizen. Because Mike Ohaco, Fermin, and his brother Miguel were noncitizens, their partnership lost their grazing permit and private land in the area where Munds Park is today.

Wickenburg would be their winter residence and where two of their five children were born. Fermin and Benancia purchased their first home in Winslow. Two children were born during the summer when the family spent their time in Winslow as that is where they had summer grazing lands. The oldest child had been born in Los Angeles. Benancia and Fermin's partners' wives would travel by train in the spring with their household goods to meet their husbands at the family ranch. The men had trailed the sheep northward earlier as it could take up to two months or longer for the trailing process, and the children

---

150   State of Arizona, "Arizona's Alien Land Act," *Revised Code of Arizona* (1928): 647–648. With this law in place, if applicants had refused to serve in the military, they could be denied citizenship.

were still in school. Mrs. Aja (Fermin Echeverria's daughter) said that when she was a child, the family would leave Winslow in October to travel to Wickenburg for her and her siblings to attend school. The sheep would graze in the nearby alfalfa fields in the winter. Stories were told of the kids visiting the sheep camps to check if the sheepherders needed anything and "taking super picnic lunches of fried chicken and hand-beaten cakes and making carvings in the aspens." Many times the whole family would go to the sheep camp for the day. Mrs. Aja said that it was a simpler time when kids made do with what was on hand to play with. Quality time was spent with the family.

Fermin had made plans to visit his homeland with his wife and children but the Depression hit. Some of his children have visited the family home in Spain. A cousin had inherited the home. A corporation was established to purchase the family's ancestral home, and in the summer of 2016, the purchase was complete. Work has been started to make cosmetic repairs to the home. It is one of the original-style homes still left in the village, and thus is important to the Basque heritage and to the Echeverria family.

Family traditions were kept alive in the food, music, and language at home. When interviewing many Basque families, the author was told that the women were responsible for activities associated with the home. They took care of the finances; early days would see them carding and spinning the wool to knit into clothes and obviously wash the clothes, cook the meals, and take care of the children. The *Sopo* was used to wash dishes in very hot water. It was made with a handcarved handle.

As children got older, they each had jobs to perform. The men mostly took care of the sheep, although the women did help in some after taking the children in the summers to the camps. Many of the younger Basque learned to play musical instruments, such as the accordion, a favorite Basque instrument. Education was also important to the families. One person related that his parents thought it was important for him to go to college so he worked for other farmers where he earned money for his college tuition. The children were taught the card game casina, which taught them how to add and subtract. The children knew how to add and subtract, multiply, and divide long before they went to school. Adults would play MUS. MUS is still an active game today as a tournament is held in Chandler.

In the early years, when there were still Basque boarding houses, parents would load their kids up and take them to the dance hall, where the parents would dance. The author was told by several Basques about these dances and the fun everyone would have. The younger kids, when they got sleepy, would just sleep along the walls of the dance hall on blankets their parents had brought. Fermin told the family that from Winslow, he would take his wife, Benancia, and Miguel and his wife, Vicencia to the dances in Flagstaff. They would ride in a Model T and if it broke down, Fermin would fix it with bailing wire. He was able to fix many pieces of equipment with

used materials. The dances were an important social outing and not to be missed! The men also played a game
similar to handball on a *perlota* court that was next to the
Basque boarding houses.

The mothers and children visited the sheep camps
during the summer months, but it was the domain of the
men. The men cooked bread in a Dutch oven buried in
the ashes of their fires. They would eat beans and mutton roasted on a spit over the fire. One article stated that
Basque were never out of work as they would always find
themselves a job, a means to take care of their families.
The Ajas and Echeverrias survived the Depression with
ingenuity and by being very frugal, something that was
instilled in them by their European parents. This trait
was passed on to their children. In all the interviews with
Buckeye Basque families and others, the one commonality
was that each said that they had been instilled with the
family values of honesty, hard work, and integrity. One
young person remembers her dad instilling in them a love
for their adopted country!

According to Irene Aja, even though Fermin's mother
told her sons not to worry about the family they left behind when they ventured forth,[151] Fermin and his brothers
sent money home to buy more land to add to the family's
holdings, paid for the modernization of the bathroom and
kitchen in their ancestral home, and purchased a tractor
for village use in 1955, for they were still harvesting their
crops by hand with scythes. They also sent money home

151 Personal communication with Irene Aja.

The handmade "El Sopo" was used to wash dishes at the camps with water boiled over a campfire. *Photograph courtesy of the author.*

through the years to help support their three widowed sisters' families. Medicine that was either unavailable or hard to obtain was sent to family members. As Irene Aja stated, "It was family, first, last, and always."[152]

Echeverria's Ancestral Home, Casa de Miguelico, in Spain.
Photograph courtesy of Irene Echeverria Aja.

In 1923, a new partnership was formed among four Basque immigrants: Mike Ohaco, Fermin, Jose Manterola, and Mario Jorajuria. The partnership was just called "Mike Ohaco." In the summer, when the men would trail the sheep northward, the wives of the four men would join

152 Ibid.

them, but the wives and children would travel by train after school was out and return the same way in the fall to ensure the children would return on time for the beginning of school.

Fermin and his brother Miguel's families lived together for many years in Wickenburg. Fermin's daughter, Irene, wrote that the rented house was "next door to a 'house of ill repute' and the men who would frequent the establishment next door sometimes made a mistake and knocked on our door."[153] Irene also told about her sister, Ellen, the oldest daughter of Fermin, and her cousin Josephine being taught arithmetic by her mother before they went to school so that when they did enter school, they were ahead of the other students in knowing their addition, subtraction, and multiplication tables. Education was important to the families, and when the father was not at home, the mother would make sure the children were educated. Fermin finally bought a home in Wickenburg in 1926, and today, the house is on the National Historical Register. It has been converted into a pizza parlor. Benancia had a green thumb, and her yard won her awards. She raised chickens, rabbits, and ducks that would become food for their dinner table.

In 1926, the earlier partnership was able to purchase property and a forest grazing permit in the Sitgreaves Forest south of Winslow on the Mogollon Rim under the name of Mike Ohaco as he had become a United States citizen. The ranch became known as the Tillman Ranch. When the

---

153 Ibid. Paul W. Pollock, "Fermin Echeverria," *American Biographical Encyclopedia Volume 1* (Phoenix, AZ: Paul W. Pollock, 1967), 204.

property was purchased, a log cabin was on the land. The duty of the men was to build, and build they did, for their families would be spending their summers here. The hand-hewn rustic ranch house would be without running water or indoor plumbing and electricity. The original log cabin was turned into a barn. Irene told the author more details:

> The meals were prepared on a wood stove. It was on this ranch that they and their families lived a true wilderness experience, for the forest was very thick, with ponderosa pine and spruce trees with only one way in to the ranch headquarters on a small, narrow truck trail. Their children became expert horsemen, riding into the deep canyons, switchbacking back and forth on the trails, along the way inscribing their names and works of art on the aspen trees. They would visit the sheep camps to check if the sheepherders needed supplies or if the sheep needed attention. The wives would pack super picnic lunches (fried chicken, the chicken having been raised on the ranch, and hand-beaten cakes). On any given day, all would set out on the truck trails to one of the camps to enjoy a few hours of camaraderie. In the evenings, out would come the deck of Mus cards. The youngsters would play the game their father taught them, *casina*. This card game taught them to add numbers before they even went to school.[154]

The partnership bought another ranch that year also, the Meyers Well Ranch. It was located southeast of Aguila. Living in the Wickenburg area, the partners had grazed their sheep on the desert surrounding the ranch in the winter months for years. Prior to the Taylor Grazing Act of 1933, livestock could be grazed on public lands on a

---

154 Personal communication with Irene Echeverria Aja, May 2015. Most of this came from a tribute written to her parents in 2000.

first-come, first-served basis. Now that the business owned private lands, the Meyers Well Ranch, the business now had state and BLM grazing leases and the sole right for grazing their sheep on the public lands surrounding the ranch.

The Depression hit, more children were born, and the business was able to weather it all, but changes were in the air in a number of cases in the 1930s. The partnership that had been formed earlier was incorporated into the Ohaco Sheep Company, Inc. in 1933. The four men would run the company together until 1941 when Mike Ohaco would sell his stock in the corporation for a quarter of the corporate assets. This gave him the ranch known as the Butte, which was south of Winslow. It had been purchased during the years of the corporation. The Butte had been part of the Hashknife Ranch.

Another change was the method of grazing sheep. Drought meant that the lush grasses in the desert were not available to the sheepmen, and they would need other methods to feed the sheep during the winter months. Because of the colder northern temperatures, the animals still needed to be moved to the central part of the state. Farmers who had raised alfalfa hay negotiated with the sheepmen to use their land for grazing purposes. The fields were in Casa Grande, Tonopah, and the Harquahala Valley. These fields were irrigated, and after hay cutting, the animals would be moved onto the fields. They would eat the stubble and at the same time fertilize and cultivate the fields with their small hooves. The animals would be moved from pasture to pasture.

The last change was the introduction of cattle to the ranch. The men understood that a ranch could have a healthier plant life when both sheep and cattle were grazed together. Sheep also have the advantage that seeds are carried and transferred in their wool between fields.

The 1930s and 1940s were mostly good years for the partnership, even though there were some changes. The corporation was able to buy and sell several ranches and mining claims. One property purchased would be Brown Springs, and each fall and spring it would be used for trailing sheep. A second partner, Tony Manterola, sold his stock in the corporation in 1945. Mario Jorajuria sold his interest in 1950. Mario would continue to live with Fermin's family as they considered him part of their family. That year saw Fermin buy a home in Phoenix, which allowed the two youngest daughters to attend St. Mary's Girls High School. This location also allowed Fermin to come home each evening unless it was lambing time.

Fermin and his sons, Fermin M. and Phillip, formed a family-held corporation in the early 1950s. Fermin and his sons brought a large flock of Rambouillet ewes to the Tonopah and Harquahala Valley to lamb. The Rambouillet grows fine wool used to make clothing. Until 1975, the Ohaco Sheep Company would graze their sheep on the alfalfa fields in the West Valley area of Phoenix, near Buckeye. That year they liquidated their breeding ewes. Fermin M. would leave the corporation in the late 1950s, only to return when Phillip was killed in a car accident and his father was no longer able to take care of the sheep business.

Fermin Sr. suffered a stroke in the summer of 1968 but recovered enough to be able to carry out the day-to-day business. He would continue to suffer mini strokes that finally put an end to his participation in the business. Benancia also had health issues resulting from a fall from which there were many complications. She was bedridden the last years of her life. With the help of private nurses, the family took care of the couple who had spent fifty-six years together in their own home until they passed away. Fermin Sr. passed away in 1974 and Benancia two years later.

Irene said her mom told her that her dad never failed to call out *"mi haytia,"* meaning "my darling" in Basque, when he would enter their home. Fermin was involved in his industry throughout his lifetime, serving as a board member of the AWGA, a member of the Arizona Livestock Sanitary Board, and a member of the United States Bureau of Land Management Grazing District Board. His daughter remarked to the author that he was successful in his adopted country because of his willingness to work hard, his pride and dignity, his openness to the American culture he encountered and a plain "determination to succeed."[155] This would not end the Echeverria's interest in the sheep industry as his youngest daughter, Irene, would marry into another sheep family, the Ajas.

Fermin's Son, Phillip

Phillip married Julia Aja (Aja sheep ranchers will be

---

155 Personal communication with Irene Echeverria Aja, May 2015.

discussed below), and they ran their own operation in the Harquahala Valley. "If it weren't for my two sons, I'd toss in the towel because this business is too much for any man, let alone a woman," Julia Echeverria told the *Arizona Gazette*, February 24, 1968. Julia came from another sheepherding family, the Ajas. Upon her husband's death in 1965, Julia took over the more than twelve thousand sheep of the Phillip Echeverria Sheep Company, Inc. at the age of forty. She and her sons ran the business, wintering the sheep in the Harquahala Valley until the early 1980s. Her sons, Phillip and Joseph, helped her after school. Then they sold their ewes and began to fatten feeder lambs on the alfalfa and grain fields in the valley for themselves and other sheepmen. While the newspaper stated she might have been the only woman running her own operation in the 1960s, there were other women prior to her. One of them would be Marianna Manterola. She was also part owner in three other sheep outfits. When Julia traveled to Spain, her sons were given the responsibility of taking care of the outfit, with the help of Julia's brothers.

Julia had more than ten Basques working for her. That year, she was proud that her outfit had garnished ten cents more per pound on the sale of the wool than other outfits. She attributed the increase to her wool being lighter and cleaner.

But even with the extra money per pound of wool, Julia still stated that like all sheep outfits, costs were increasing as the wool prices continued to decline. Workers imported from Basque Country, at least all of hers

Where Have all the Sheep Gone?

came from there, cost her $50 more per month per worker. Workers have to be paid even during months when they are not needed; otherwise, she would not have them when needed. Another problem facing all sheep operations, Julia stated, was the federal government allowing importation of lambs and wool from New Zealand. This was not the first time that the author had heard that government regulations were stifling the American sheep business. New Zealand's government subsidized the sheep industry and thus costs are lower to raise sheep. Today, the government still subsidizes their industry making for importation of feeder lambs cheap to import to other countries, especially the United States. Julia and her sons sold their interest in the corporation in the early 1970s. She ran her own outfit until she sold out in the late 1980s, thus ending another sheepherder family.

Mike Echeverria, Son of Miguel

Mike was the oldest son of Miguel and Vicencia and followed in his father's footsteps. Mike married a daughter of Cruz Eraso, a Basque sheepherding family. He had five children. His son Tom said his father was a trader. In the kitchen, his father had a bulletin board divided into twelve sections, and in each one was information on a hotel in which their father could be found in Texas. Mike traveled a great deal and was continuously receiving calls for him to buy or sell sheep. He averaged 200,000 buys or sells a year. The home phone was considered a business line, giving the children thirty seconds to use the phone. If Mike would call home and the phone was busy (the line at this time was a party line), the dad would have the operator cut in on the phone call. Tom said, "You didn't want to be the one on the line when this happened."[156]

The Ajas

The Ajas are very proud of their Basque heritage and being a part of the sheep business in Arizona. Irene Echeverria Aja was instrumental in giving the author contact information for sheep families within the state. She had been compiling information about the sheep business in the state and especially the role her family, the Echeverrias, and her married family, the Ajas, played in the development of the industry. Irene has collected information on families who called the Buckeye Valley home. She was an advocate for the industry back in the 1980s and 1990s. It was very evident that she was and still is proud of her heritage and the part the Basque played in the economic development of Arizona. The story of the Ajas cannot be separated from the Echeverrias, as Irene's brother (Philip discussed above) married an Aja, and she herself would marry one also.

The Ajas, Manuel and Fred, arrived in the United States in 1919. By shoveling coal into a boiler of a ship, Manuel made his way to America in 1920. Like many other people in Europe and especially in the Basque area, times were hard and people were starving. He set out to make his fortune in America. Manuel met his Spanish-born wife, Matilde Cuevas, in America, and they

---

156 Personal communication with Tom Echeverria, October 7, 2016.

married in 1926. Manuel Aja, his wife's brother, Basilio Cuevas, and her uncle, Manuel Cuevas, and Federico (Fred) Aja, brought two thousand sheep into the West Valley in 1934. Manuel, Basilio, and Manual Cuevas worked for other sheep outfits, and Fred ran their sheep. Circa 1940, Manuel Sr. and his partners bought a ranch north of Williams, which had a trail right from Ash Fork. Around 1939, Fred sold his interest to his partners and bought a cattle ranch south of Winslow, where he ran cattle and sold summer pasture to several sheep outfits and allowed family members to graze their sheep.

Manuel and his wife had six children: Manuel Jr., Robert, Julia, Benny, Basilio (Bas), and Virginia. Manuel bought the sheep from his partners in 1945, and he was the only proprietor until 1950 when his three sons, Manuel Jr., Basilio, and Robert, joined him. Manuel Jr. and Bas were also partners with Tomas Cuesta in the 1940s. In 1950, the two brothers bought out Cuesta and joined their sheep outfit with their father's. Bas and Manuel Jr. put two thousand sheep with their father's two thousand sheep, and the father gave a thousand of those sheep to Robert. The outfit was called the Aja Sheep Company.

In 1958, Manuel Sr. died from injuries he suffered in a car accident. Even though a family corporation was formed in 1960s with his sons, Robert, Manuel, Benny, and their mother, Matilde, they would be out of the sheep business within the next ten years. Robert and his wife, Doris, sold out and moved to Idaho in 1971. Manuel

Jr. sold his interest in 1967, and he and his wife, Dona, moved to Colorado in 1984.

Bas courted a girl he met at the Wool Growers Association party in Flagstaff. Bas and Irene married in May 1949 and moved to Buckeye. Irene told the author it was a trailer without many amenities, and she was glad when they built their home in 1951 on property given by Bas's father. The couple had six children: Melanie, Bas, Roy, Rachel, Gigette, and Christine. Roy, Bas, and Alberto Uriz ran sheep in the late 1970s in the Tonopah area. Roy sold out in 1988 and Bas in 1990 when he went to work as executive director of Arizona Cattle Feeders and began to run cattle. Melanie is the only family member to still have sheep. She raises St. Croix sheep, a hairless sheep, for their meat, which is sold to Jewish markets. The St. Croix breed is listed as endangered.

The Ajas grazed their sheep on the local alfalfa and grain fields in the winter and trucked the sheep northward in the summer. In 1961, after his father's death, Bas separated from his brothers and ran sheep on his own. The couple leased several ranches, the Wine Glass, the Del Rio, 11 Lakes, Juniper Wood, the Gold Trap, the Nagel, and the Fico, in the northern portion of the state until they purchased the Black Rock Ranch in the late 1970s. They had rented it for four years prior, near Joseph City. His daughter wrote that he worked hard to buy a fifty-thousand-acre ranch. The land was adjacent to the Navajo Indian Reservation. Irene recently showed the author one of many blankets that the Navajo women made for her when they lived on the ranch. The ranch covered two townships, giving them plenty of land to graze over seven thousand ewes and some cattle. He ran over ten thousand sheep for other sheepmen.

Bas's children spoke highly of their dad and the work ethic he instilled in them. Gigette stated that he was an environmentalist long before it was "politically correct" to be one, "that if you did not take care of the land, it would not take care of your animals, and in the end, not take care of you."[157]

Bas continued to run a Rambouillet breeding ewe operation until 2005. Then the Aja Sheep Company sold their breeder enterprise and turned to fattening feeder lambs for market, themselves, and other sheepmen. When regulations created too many problems, costs increased, and it became difficult to find good help, Bas sold his operation in 2005 to Mr. Etchagary, a Basque in California. The ranch was sold in 2011.

Bas passed away in April 2014. Irene continues to live in the family homestead with many neighbors and busy streets on two sides of the house. She has seen many changes in the sheep industry in her life, the greatest of which is the dwindling numbers of sheep raised in the state and the number of outfits.

The Ohacos[158]

Michel and Louise Aristoy Ohaco were both born in Basque Country. What is known about their involvement

---

157 Memorial written and read at Bas's funeral by his daughter, Gigette.
158 Paul W. Pollock, "Michel Joseph O'Haco," *American Biographical Encyclopedia Volume 1* (Phoenix, AZ: Paul W. Pollock, 1967), 78.

in the sheep business actually comes from Michel's involvement with other sheep families.

Never a sheepman himself, Michel grew up in a Basque family in the sheep business. Michel Joseph Ohaco was born in 1920 in Phoenix. His parents were in the sheep business, and he spent his early years traveling between the northern area around Flagstaff and the Wickenburg area. His elementary school years were split between the two cities until teachers advised the family to have him remain in the same school for the seventh and eighth grades, which were then attended in Wickenburg. Michel was an exceptional student and athlete. He lettered in many sports and continued when he went to the University of Arizona. There he majored in agriculture and earned a commission with the army ROTC. After graduating, he completed the rest of his military training before being sent overseas to fight in World War II. He was severely injured, and while recovering in a hospital in California, he met his future wife, Teresa Savinsky. Michel returned to Arizona in 1948 just as the last of the family sheep interests were sold. Michel took over the family ranch, Chevelon Butte, and from then on, it was a cattle ranch.

The Alemans

John M. Aleman, born in 1894 in the Basque area of Spain, arrived in Winslow on March 3, 1914, after a boyhood friend, Juan Bicondon, who had preceded him by one year, wrote of the many great things about Arizona. Juan wrote of "a land full of golden opportunities, knee-high grass where sheep and cattle grazed as far as the eye

St. Croix Sheep. This breed came from the island of St. Croix, Caribbean. The Spaniards brought the sheep to the island and it adapted well to the weather. The sheep are parasite resistant and adapted well to the heat and shed their hair. *Photograph courtesy of Melanie Aja Lanford.*

could see, and cowboys riding horseback with forty-fives on their hips and Winchesters on their saddles"[159] His friend said that you could walk for miles and not see anyone else. John had to come to see for himself. A few days later, he began to work for Spellmire and Lyons on their sheep ranch. He worked as a sheep tender or cook for them. John's wife wrote that "he quickly learned to cook over an open fire, baking Basque bread in Dutch ovens, cooking pinto beans underground, and roasting lamb over hot coals."[160] As tender, he also had other jobs. He was given fourteen burros for packing the camp's food and personal belongings. The shepherd's personal belongings and food for the camp were kept in chuck boxes, and two of each were strapped to the burros. Five-gallon water barrels were also carried by the burros for water used in the camp. On top the boxes, bedrolls and the tent were added. Last, a canvas cover was put over the supplies. Each day John had to take the burros for water and replenish the water in the barrels. Another responsibility was taking eight of the burros back to the ranch to replenish the camp supplies and deliver any news. This occurred twice a month. John's wife, Frances, wrote after his death that he always hoped that there would be mail or even an old newspaper he could read. "Life was very lonely for a young man who loved people and life," she wrote. [161]

---

159 A copy of the family history written by Francis Aleman was given to the author by Francis' daughter-in-law.
160 Ibid.
161 Ibid.

He continued to work for them for several months when he left to work for the Frisco Mountain Sheep Company, a partnership between M. I. Powers and Pete Espil Sr. This outfit was headquartered in the Flagstaff area. In October 1914, John found himself helping trail their herd of sheep to the winter grazing around Wickenburg, and there the lambs would be born in the spring. Frances said that John told her that he would never forget his first winter in the desert area of Wickenburg. She wrote, "The rains came in floods, all wood was soaked, and he could not make a fire for days at a time. If by chance he was able to get a fire going and had the food cooking, there would come another downpour, and everything would be ruined again. At night he would have to cut bushes and lay them out like a bed two or three feet high and put the bed rolls on top of this because the ground was covered with water."[162]

He worked for them two years in the Wickenburg area, being promoted to herder. Three years later, he was promoted to foreman when he moved on to work for Tom Pollock on U Bar at Clear Creek in 1924 and then Mike Ohaco's sheep company in 1925.

In 1926, John and Miguel Echeverria bought the Cross Mountain Sheep Company from Campbell and Francis, which was located at Seligman, Arizona. They ran the company together until 1941, when he sold his interest to Miguel's son, Miguel, Jr. This was not the only sheep company John had an interest in. With Joseph and Leon Pouquette, the three purchased Lou Charlebois Sheep

162 Ibid.

Company and the Boulin Ranch in 1935. They changed the name of the sheep company to Pouquette-Aleman Sheep Company. This partnership took place a year before he married Frances Hendrix in 1935. Frances, on a visit to her father's Box H Ranch in Concho in 1929, met John when he first began to rent her dad's ranch to graze his sheep in the winter. In 1945, the company was dissolved, and Joseph Pouquette took a quarter of the sheep. Leon had passed away in 1943, but the partnership with his widow and John continued until 1952, when John gave a third of the sheep to Leon's widow, Amelie. With the help of her son, Felix, they ran the sheep. Also, in 1952, John began his own operation under the name of John Aleman Sheep Company.

It was mainly the men's job to herd the sheep, and the women would take care of the financial aspects of the business. The women would purchase the food for the herders as they tended to the sheep in the summer camps. She purchased medical supplies for both the men and sheep. If a herder needed a doctor or dentist, the wool grower's wife acted as chauffeur. When schoolchildren would visit, the women would be responsible for showing the children the activities. Mrs. Aleman gave cooking lessons to school children who were interested in cooking lamb. She stated, "Many people don't know how good lamb is, because they've never tasted properly cooked lamb—or maybe it wasn't lamb at all."[163] Several families interviewed told the author the same story.

---

163 Nourse, 38.

Early in the couple's marriage, Frances was very active in all aspects of the sheep industry, besides the bookkeeping part. She helped her husband in all facets of taking care of the sheep. She would ride with him when he trailed sheep from Williams to Wickenburg. "I slept on a pelican bed (a canvas cot) many a night in those years."[164] They would ride ahead of the sheep and herders, looking for waterholes. Their job was having lunch ready when they met the herders in the middle of the day. Afterward, John and Frances would begin to find a good camping place to spend the night. It was not uncommon for Frances and John to ride ten to twelve miles each day. Waterholes were her bathtub except for the time she saw a snake slither into the rocks around the waterhole.[165] Wives helped the men during dipping and shearing times too. They would prepare the food for the men that would include growers, sheep inspectors, forest rangers, shearers, and wool buyers, depending on the activity.

The sheep trail the Alemans used was Old Bear Springs Driveway. It began in Williams and ended near Congress Junction, west of Wickenburg. When the Alemans used the trail, the lambs were born in February. Drought changed the time of year that the lambs were born in the desert area to November and December. In later years, the pregnant ewes would be shipped by rail, but the railroad discontinued shipping livestock in 1970. In the last years of their sheep operation, sheep were

---

164 Ibid., 36.
165 Ibid., 36.

moved by four-decker stock trucks to their summer grazing area in the Williams area. From Williams, the sheep were trailed three days out to the Aleman's ranch Squaw Mountain Ranch.

Jose Aleman with sheep. *Photograph courtesy of Kathy Aleman.*

Frances was known as the "First Lady of the Sheep Growing Clan."[166] The Make It Yourself with Wool contest was cofounded by Frances Aleman and Ora Chipman in 1947. The Women's Auxiliary of the National Wool Grow-

---

166 Ibid., 36.

ers Association sponsored the contest. At this time, the AWGA did not have a Women's Auxiliary. It would take eleven years for Frances and other wives whose husbands belonged to the AWGA to organize one. As the charter president, Frances served for twelve consecutive years. She went on to be elected to the national association and became its president after two years, serving for two years. She was also National Press Correspondent for the organization. In 1968, she was elected Woman of the Year for her business and civic activities.

The first contest for the Make It Yourself with Wool involved twelve western states with only women and girls being allowed to compete. Each state held its own competition and winners went to compete on the national level. Today, both males and females of all ages may enter a project made with wool in the competition in many categories. The fabric from which the project is made must be 60 percent or greater of wool. There are six major categories of competition: Preteen, Junior, Senior, Adult, Made for Others, and the College Fashion or Apparel Design Category.[167] All but twelve states have this organization, with Virginia being the newest state to form an association. Contests are held each year in each of those states. Sponsors of the contest are the American Wool Council, the American Sheep Industry, and American Sheep Industry Women.

Frances learned to cook Basque dishes. One of her titles was "boss lady" of the Lamb and Wool Fall Fiesta

---

167 Personal communication with Rali Burleson, director, Arizona Make it with Wool. October 18, 2016.

Week. This event would be held in large shopping centers in the Phoenix area. It introduced thousands of people to the proper way lamb should be cooked. Frances would use the opportunity to explain the economic importance of the sheep industry.

Stories told by the sheepherders and their families showed the difficulties these men faced trying to raise sheep and provide for their families. One story told by Mrs. Aleman happened prior to their marriage. Leaving the nine thousand sheep in capable hands, the young couple planned a day outing to Holbrook, but John decided to see how the sheep were watering at the Little Colorado River. Disaster had struck as the sheep were caught in the quicksand. The couple rushed into the water, trying to rescue all the sheep, but in the end, many sheep were lost that day. Many sheepherders lost everything more than once, and Aleman was one of them. During the Great Depression, he had put his sheep on railcars to send to Kansas City. He had never gone to oversee the operation and wanted to go that year. He went into the bank to withdrawal some cash for the trip only to find that the banks were closed; there was no money. John told the banker he needed to go to Kansas City with his sheep, but the only thing the banker could do for him was give him $5 of his own money. Aleman went and stretched that money with very leaning eating, as he later recounted. This was the first of his bankruptcies as he could only sell his sheep for between $2 and $3 when they should have gone higher. He lost more when he decided to hold out for more money, but that only gave him a large sheep feed

Frances Aleman with Robert Pouquette.
Photograph courtesy of Kathy Aleman.

bill. It was said that he lost $20,000 on that transaction. The commission firm gave him money to return to Arizona, and the couple began again.

Frances wrote a somewhat humorous story her husband told about the hostilities between the cattlemen and sheepmen. It was believed that cattle would not eat after the sheep had been in an area, and the sheep would not drink from the same water hole as the cattle. This has been proven false as both sheep and cattle would later be run together.

> John loved to tell how many nights, after the cowboys left the range for the ranch, the sheepmen would drive their sheep to the water tanks to drink, and when they finished would herd them away as quickly as possible. They would then cut large branches of cedar bows and wipe away the sheep tracks, working most of the night. When morning came, the cowboys would return to the water tanks to look after the cattle and to see if the sheep had been there, but there were no telltale tracks, and the sheep were far away."[168]

The Alemans began to move their sheep in the later years by rail from the Salt River Valley to Williams and then only trail them three days to their ranch, Squaw Mountain Ranch, which they had purchased around 1935.

But it was not all work for the men. After school was out, the wives and children would come to visit. John would be able to come to see his family more often with the family close on the ranch. The children would visit their father and eat delicious meals at cookouts. Frances

---

168 A copy of the family history written by Francis Aleman was given to the author by Francis' daughter-in-law.

wrote about the food and how it was prepared, "Beans baked all night underground and barbecue lamb are staples for such occasions, with crusty Basque bread baked in the hot coals."[169]

The Aleman's son, John Jr., became a junior partner in the business in 1967, and the name of the company became Aleman and Son Sheep Company. Just like Frances, Claudia, their son's wife, helped with many aspects of the business. Before the sheep are returned to the valley, they are on dry pastures up north. When the sheep are brought into the winter grazing alfalfa fields in the Salt River Valley, many have a tendency to overeat and bloat, and they will need shots. Claudia learned to do this very well.

John had many passions besides his sheep: his grandchildren and cooking. John loved to cook, and he was especially fond of *Bildocha Irrisaikin* or lamb with rice. He could make Basque bread and biscuits in a Dutch oven but was even prouder when his only granddaughter learned to cook these foods. John played an accordion and sang many Basque songs for his grandchildren, John III and Heidi Jo. They loved to hear him play and to dance. John III learned to work with the sheep, just like his dad and granddad. He made the hobbles used to tie twin sheep together.

Before John's death, he had been an active member of the AWGA for fifty years and a director for twenty. He served as a director of the Kaibab Advisory Board

---

169 Ibid.

in Williams for many years. John Sr. died in 1972 at the age of seventy-nine, and his wife died in 1983 at the age of seventy-two. John and Claudia divorced in 1977, and John remarried in 1980 to Kathy Wolfswinkel. Kathy continued Frances's legacy by becoming a Wool Growers Auxiliary president. To this day, Kathy continues to help sponsor the Make It with Wool state contest, and the junior and senior winners go on to the national competition. As Kathy wrote, "This competition is still going strong after sixty-nine years, promoting wool fabric."[170]

Just like other Basque, the Alemans have visited the homestead in Basque in Spain several times. The homestead plays an important role in their life, and time spent in the area where they grew up is cherished. John's younger brother inherited the home that has seen five generations of Alemans. The Aleman's first visit to Spain occurred in 1948, thirty-four years after John left. Their son John Jr. accompanied them. Frances wrote of her recollections of the trip in great detail, describing the home and the three levels, each with a different intent.

The first floor was used for the livestock, equipment, and some household chores. To one side of this floor was a huge oven that was used twice a week to bake the bread. The washing was done in a cement square water trough with a slanting cement slab with grooves, the washing board. John got to sleep in the bedroom of his great-grandfather's, located on the second floor of the three-story house. The living quarters had a large living and dining

---

170 Personal communication with Kathy Aleman, September 23, 2016

John Aleman, Sr., at sheep camp. Boxes that carried their cooking supplies and gear were now the work tables at the camp. *Photograph courtesy of Kathy Aleman.*

area and a very big kitchen. In the kitchen was a wood stove, fireplace, cement sink with only cold running water, large cupboard, table large enough to seat twelve people, and a pantry. Besides the bedroom of his great-grandfather, there were five others. The floor of the home was made out of three-inch thick hardwood and polished to look like a mirror! The third floor of the home had two purposes: one to store food for both animal and humans: hay, corn, potatoes, beans, onions, garlic, fruit, nuts, and whatever else the family had grown or collected. The second was to dry clothes on the balcony. Frances remarked that this was the summer she really became a Basque as she had learned Basque culture and way of life.

A cute story told about John Jr. had him wondering why the cowbells were full of straw. He was told no one knew, and after pondering the situation for a little while concluded it was to allow the cows to sleep at night and the bell clap would not keep them awake. The real reason was to allow the family to sleep!

While the Alemans visited with family, they also spent time meeting with the families of the herders they employed in Arizona. They visited families of herders who were in the employment of other Arizonans, exchanging letters and pictures. John also met and arranged for two Basque men to come and work for him in the States, Jean Arriaga and Tony Echandi. How these Basque herders and others came into the United States will be discussed in the last chapter, as it was one of the reasons for the decline of the sheep ranchers in Arizona.

John Aleman, Jr. was responsible for taking care of the burros at an early age. *Photograph courtesy of Kathy Aleman.*

The Manterolas

Jose Antonio (Tony) Manterola arrived in the United States in 1907 at the age of seventeen. He was born a Spanish Basque in Sumbilla, Navarra. He was one of six children. When Tony was a small child, his father died, and when he could, he left for America to seek his fortune. His first herding job was in Van Horn, Texas. His daughter, Carmen, said that he enjoyed telling his family of the first days in America and of his cooking skills. His boss in Van Horn took him out to a range, where he was to stay alone, herding a band of sheep and tending his own camp. His cooking skills were few, and his daughter said she remembered him telling that the first biscuits he baked were so hard that not even the burros would eat them. In later years, he was a very good cook. After spending three years in Texas, he came to Arizona, where he had friends working for other large sheep outfits. He went to work for H. B. Kelly, owner of the Kelly Sheep Company. It was said that the methods employed here to raise lambs weren't much different than those practiced in his homeland.[171] Embach wrote that he met Manterola for the first time when he and his cattlemen rode into the sheepman's camp and were offered hot chili. They remained friends throughout the years.[172]

He also worked for Dr. Ralph O. Raymond, a doctor who had sheep in Arizona and New Mexico. Tony bought out a half interest in their business, which consisted of two thousand sheep, seventy-four rams, and ten burros. Tony operated this business for a number of years and later sold his interest to Bernardo Bidegain and Jean Arriaga. He then went to Calipatria, California, with a friend, where they had dreams of making a fortune in the cattle business. This venture proved unsuccessful, and he returned to work as a sheepherder for Hart Sheep Company, headquartered in Flagstaff.

In 1922, the Ohaco Sheep Company was formed with Tony Manterola, Fermin Echeverria, Mario Jorajuria, and Mike Ohaco as equal partners. The four Basque men

---

171 Harry B. Embach, "Jose Antonio Manterola—A Real Sheepman," *Arizona Stockman* (May 1950), 34.

172 Ibid.

successfully operated their business, said to have been the largest sheep company in the state. Tony would trail the sheep from the Wickenburg winter grazing range to the summer grazing area near Winslow. He also began to take ewes to lamb in the irrigated fields of what was then the west valley area of Glendale and Litchfield.

In 1932, Tony Manterola and Marianne Etchart, who had emigrated at the age of seventeen from the French Basque side of the Pyrenees, married in Wickenburg. Fermin Echeverria and his wife, Benancia, were witnesses. The two families would grow close in the next few years. Marianne had arrived in Flagstaff in 1928, where her cousins lived, and went to work as a housekeeper for the Tom McCullough family, a pioneer Flagstaff family. She later worked in Phoenix for the Dr. Matanovich family as a nanny for their children until her marriage in 1932. She had been born in Argentina in 1909 when her young parents had emigrated to Argentina from La Madeleine, Basses Pyrenees, France. The family returned to their native country when she was three years old because of difficulties of living in the Pampas. Her father would commute between the two countries as he continued to raise sheep in both.

During their marriage, they had four children: Sylvia, Carmen, Marie, and a son, Jose Jr. The family first lived in the Glendale and Peoria areas. They would spend their summers in the cool North Country, where the sheep were grazed and bred. Carmen went on to explain about her father's business: "My dad ran a split operation lambing ewes in November for an Easter market and in

Bloody Basin for a July market."[173] Twenty-five hundred ewes were wintered in the Casa Grande area, and the lambs went to the Easter market. The same number of ewes lambed in February, making them ready for shipment in the summer.

The Tillman Ranch, located about sixty miles south of Winslow, was their home during the summer. They lived in a log cabin and hauled water in lard buckets from the well. Carmen remembers her mother cooked for the ranch hands as well as cared for her family. Carmen told the author, "We spent all summer at the ranch, only coming to town to prepare for school in September. Once a week, my Dad and Fermin Echeverria would go to Winslow for supplies and mail. When we saw the truck coming back through the meadow, we would run to meet it, as our treat was candy and fresh fruit."[174]

In 1945, Tony sold his interest in the O'Haco Sheep Company and bought the Flagstaff Sheep Company from Dr. R. O. Raymond. The Manterola Sheep Company was begun that year. He then acquired the Woody Mountain Allotment and the Mooney Mountain Allotment, the Tonto Forest Allotment and the Bloody Basin Allotments, which were needed to graze his sheep upon each summer. He also got the bridge crossing the Verde River that Frank Auza and a number of Basque built. Over the years, he acquired various forest and private lands on the Kaibab Forest adjoining his operations.

---

173 Personal interview with Carmen Auza, May 2016.
174 Ibid.

Tony and Marianne Manterola.
Photograph courtesy of the Manterola Family.

In 1945, the Manterolas moved to a new ranch located east of Williams in Garland Prairie. Their children remembered the good times they had while visiting the ranch. Carmen told the author:

> There are many fond memories of life on the ranch for us Manterola children. Cooking on a wood stove, carrying buckets of water from the well, chopping wood, and heating water in a tub for washing clothes on a washboard were all everyday chores. Mother and my older sister, Sylvia, washed clothes on a washboard. Marie and I were responsible for pumping the well water with the hand pump, and then we hauled the hot and cold water for them. We had to replenish the fire with wood to keep the water hot. In later years, the washboard was replaced with a Maytag gasoline-operated washer that my Dad would start before leaving, as it was very hard to start.

Carmen said she remembers riding the burros and swimming in the lily pond in Sycamore Canyon. She told of how she and her sisters learned to sew on a Singer treadle machine and made some of their school clothes during summer vacations. They knit with huge nails while they played in the swings hung from the trees. They also learned to drive on the country roads.

Both Tony and Marianna were active during their lifetimes. Tony served four terms on the board of directors for the AWGA. Governor Dan Garvey appointed him to serve on the Sheep Sanitary Board. When it was time for the AWGA annual meeting, Tony and Marianna hosted it at their Aspen Springs Ranch, near Flagstaff. Hundreds of people, many of them dignitaries, would attend the event, and barbeque lamb was always served. Marianna was an active member in the AWGA and its auxiliary, Arizona Cattle Growers Association, and was recognized as an Arizona Pioneer Rancher at the National Livestock show since age seventy-five.

When Tony died in 1956, Marianne and the oldest daughter, Sylvia, ran the sheep outfit along with Frank Auza. Frank's son, Joe, took over for his father when Frank went off on his own. Joe Auza was married to Tony's daughter Carmen. Tony Manterola's son, Joe, took over the outfit from Joe Auza when he became old enough. It became a corporation in 1968, and both cattle and sheep were raised. Marianna was still involved in the sheep business when she passed the century mark in 2009. She would help where she was needed during shipping, shearing, and cooking for the sheepherders. She lived until she was 102. Today the outfit is run by Tony's children, Sylvia and Joe.

Joe remembered when he was little. His mother spoke four languages, Basque, French, English, and Spanish. The father told the mother she was not to speak another word of English to him until he could speak Spanish. He was six years old. He learned quickly from his mother. He also gave his view on the state of the industry through the years.[175] Joe stated that in the forties and fifties, one hundred and fifty thousand sheep grazed in the Salt River Valley between Chandler and Litchfield. Because of

---

175 Oral History Interview with Joe Manterola, June 18, 2008. Colorado Plateau Archives, Northern Arizona University Cline Library.

housing developments, the Manterolas began to graze their flocks in the winter between Needles, California, and Laughlin, Nevada. Later they run quite a few feeder lambs in the Casa Grande and Coolidge areas. Housing developments encroached here, along with dairies. The alfalfa fields that once were used for grazing the sheep in the winter are now cropped all year and feed the cows. The price of hay means they can't lease the land. Joe saw the end of the railroad being used by the sheep and cattle in the late 1960s. Trucks were introduced and livestock corrals were created. Where the Flagstaff Sam's Club is today, there was a large staging area for loading the animals to move southward. Joe also discussed why so many ranchers sold out in the 1950s and 1960s. Deeded land meant paying taxes, and the price was high compared to leasing Forest Service land. Many people, including his dad, bought these ranches at $30 to $50 an acre. Joe said that most people were ready to move into town after trying to winter livestock and baling hay and other feed for the animals. They were happy to get any price for it. Drought also forced many of them to sell, but Joe feels that the land is in better condition than when there were all those ranches.

When the author met Joe during tagging season in 2016, he was full of stories of life with the sheep. I asked how he was introduced to the sheep business by his father since he was the only boy and expected to take over one day. He told me that his father dumped him out with a herder and told the herder to teach him the business. He thought his dad was the meanest person in the world to do this to him, but he soon learned his dad was serious about him learning the business. On this occasion, Joe spoke fondly of his sheep and the business he inherited. When asked why he stayed in the business, he said it was the only one he knew. Joe has no son (he was killed in a car accident) to carry on the business; he does not know what will happen to his sheep. He talked about the hardships of being away from his family and other problems, such as rising costs. Through spending time with Joe, it is obvious he loves the business even with all of its headaches.

The Auzas

The Auzas have been involved in the sheep business for 101 years when the father, Frank, emigrated in 1915 at the tender age of ten. Frank's father had emigrated a year earlier, and the mother followed later, bringing the children. Frank's parents died when he was a teenager, his father from an ulcer when he was fifteen and his mother of pneumonia when he was eighteen. He was already a hardworking man by this age, for at age eight he worked for people in Spain who bought and sold cows and sheep. He was paid $15 for the two years of work and was given food. When Frank arrived in America, he only spoke Basque and used sign language to speak before learning both Spanish and English. At age fourteen, Frank began taking care of cows, getting paid $15 a month, but soon switched jobs and started to make $3 a day packing grass. When the family moved to Flagstaff, Frank went to work for Pierre Espil and continued for about eight years. Sheep became a part of his life when he went to work for Dr. Raymond, herding sheep at Howard Lake. In 1933, Frank married Elsie, and they had eight children. Frank helped run the Manterola's sheep company after Tony Manterola died. He ran the company with his widow and two of their children. In 1959, Frank bought Dr. Raymond's sheep company and continued to run that outfit. Frank's brother, Joe, then helped the Manterolas. Until 1988, four of the children, Tine, Frank, Pete, and John, ran sheep together in the Yuma area. Tine, his wife, and their children raise sheep in Imperial Valley area of California and thus are not part of the Arizona story except to mention that Tine was born in Arizona into a sheep-raising family.

Carmen Manterola married into the Auza family in 1962. They had two children, Joe Jr. and Yvette. Joe Jr. also runs sheep today under Joe Auza Jr. Sheep Company. The Auzas winter their sheep in Casa Grande while Manterolas winter theirs in Mohave Valley.

Sometimes the only available information about any one particular sheepman or his outfit comes from records kept by the AWGA.[176] From these files, it is possible to sometimes determine how many sheep some of the outfits ran and where their sheep grazed on public lands. From the 1918 dues, we know that there were forty-nine sheep outfits. Without the amount due per sheep, it is impossible to determine how many sheep any one outfit ran and thus no total number of sheep run at this time for those who

---

176 Arizona Wool Growers Association, NAU.MS.233, Cline Library. Special Collections and Archives Dept. Files accessed June to August 2016.

belonged to the association. Letters to the AWGA from the United States Department of Agriculture detailed the number of sheep that each permittee was allowed to cross on the Bear Springs Driveway, Mud Tank Government Gap Driveway, and Beaverhead-Grief Hill Driveway in 1930. There were thirty-three outfits with a total of 50,515 sheep using the Bear Springs, 57,294 sheep on the Mud Tank Government Gap, and 63,352 on Beaverhead-Grief Hill.

Three generations of the Auza's involved in the sheep industry.
Photograph courtesy of the author.

The number of outfits between 1944 and 1961, based on the total pounds of wool each outfit shipped through the AWGA, was sixty-three. Collectively, they had shipped over six and a half million pounds of wool at a price of nearly $600,000. The AWGA also would write letters to the Selective Service System asking for deferment for sheepherders and owner's sons, and many times, it would state that the family needed an experienced herder because of the number of sheep involved. There may be no other information on that outfit. In three different letters to the Selective Service System written on behave of or by employees in the service of Zack Bezunartea, it is known that Bezunartea had approximately six thousand head of sheep. These letters are dated from late 1950 to late 1952. In all three letters, the same number of sheep was reported. We know that Bezunartea was not running sheep in 1918 as his name was not on the dues list. The name does not appear on the 1930 letter mentioned above. It is possible he could have been running sheep and not been a member of the AWGA.

Another such letter to the Selective Service System dealt with the son of Bernardo Bidegain. His name appears on the 1918 dues lists but not on the 1950 list.

In the annual meeting for the AWGA, resolutions were passed for deceased ranchers. One piece of paper listed thirteen sheepmen who had passed from the thirty-eighth annual meeting up to the seventy-first annual meeting. It was found in the folder for the Francis Sheep Company even though there was no obvious connection with Francis Sheep Company. The importance of these letters is that they tell the story of the many men who ran sheep in the state. The list stating those who had died tell the researcher why no other information can be found for certain men or the sheep company. Sometimes this was the first time that a name associated with the industry came to light. In some files, the only piece of paper would be an announcement of date of the death and where the rancher ran their sheep, as in the case of Rafael Sarabia. These records are invaluable in their ability to tell the story of the sheep industry.

Jean Arriaga[177]

While hardly any information is known about Jean, a French Basque, it is known that Jean initially came to work for Jean Aleman in either the fall of 1948 or the spring of 1949. Mr. Arriaga was a member of the AWGA from 1948 to 1979. From 1944 to 1961, his total wool weight was 12,096 pounds.

The Elorgas: Joe and Alejandro[178]

The two brothers were born in Errazu, Navarra, Spain. Joe was born in 1885 and came to the United States when he was fifteen. His brother was eight years his junior and arrived in Arizona in 1911. Originally, Joe partnered in the sheep business with E. A. Sawyer and Ed Woolf in the Sawyer Sheep Company. He sold his interest in 1916 to Ysidor (Ysi) Otondo, and the sheep outfit name changed to Sawyer-Otondo Sheep Company. Joe became a naturalized citizen in 1921, and the next year joined the Wool Growers Association. A year after that, Joe and his brother, Alejandro, along with Frank Er-

177 Arizona Wool Growers Association, NAU.MS.233 Series 1.1.9, Cline Library. Special Collections and Archives Dept. Files accessed June to August 2016.
178 Ibid. Series 1.2.1 and Series 1.2.2

ramuzpe Sr., bought the Sancet and Fisher Sheep Outfit. When Joe became a citizen, he could get permits to graze on public lands. That same year, Joe became a member of the AWGA. In another letter dated 1935, this time from the Forest Service to Joe, the Forest Service explained that Joe and another outfit by the name of Ryan would have to move their grazing to the west end of the Sitgreaves Forest. The Forest Service was inquiring as to which of the two allotments available to the outfits, Rock Tank Allotment and Willow Creek Allotment, would Joe prefer for that coming summer grazing.

The two brothers sold their interest to Frank and went into business for themselves. It is known that in 1928, Frank Erramuzpe, Sr. sold them his outfit since he was returning to Spain for a visit. In 1935, a letter written to Joe from the Wool Growers Association explained the cost to each sheep outfit running sheep in the Queen Creek District for the cost of hiring a man to poison coyotes. Joe's charges were based on him running two thousand sheep. Joe was still in the sheep business as late as October 24, 1940, because he wrote a letter explaining why his employee, Jean Baptista Arambel, did not register for the draft. The letter states that Jean was on the sheep trail and "could not abandon his employer's sheep to the mercy of predatory animals or the elements." The last piece of information available about Joe was the Wool Growers Association recording that Joe passed away in April 28, 1941, just six months after he wrote the letter for Arambel.

When it was time for the sheepherder to register for

the draft, many times they would be on the trail with their own sheep, but most of the time they were working for someone else. Many letters were found in the AWGA files where an employer wrote to the Selective Service System detailing why a particular person within their employment had not registered on the requested day. The herder took this letter to the Selective Service System usually within just a few days of the original required registration date. There was no letter found in the files indicating that the Selective Service System found this to be unacceptable. Many letters were written not only by the employers of herders but also by the AWGA when a man was called to serve especially during World War II. Sometimes these draftees were sons of the sheep owner and were very important to the operation of the sheep outfit. Letters were written for exemptions for them or once they had gone into the service to ask for a discharge. Getting good workers and keeping them was a problem for all the wool men. It is one of the reasons many men sold out in later years.

The Erramuzpes: Frank Sr. and Jr. [179]

The Erramuzpes have been involved in the sheep business since the early 1900s. Born in Erazu, Spain, and into a Basque family, Frank Sr. came to the United States in 1912 when he was twenty-one. His sister, Marceline, was already in California working as a maid in the home of a Basque sheepherder whose ranch was near the La

179 Personal interview with Frank Jr. and his wife, Marcie Erramuzpe, September 21, 2016.

Puenta/Fullerton area. Frank Sr. went to work for his sister's employer as a ranch hand. Marceline married Isidore Otondo, who lived in Arizona and owned the Otondo Sheep Company. Shortly thereafter, Frank arrived in Arizona and went to work for Otondo. It is unknown how many years he worked for Otondo or the exact date he purchased his own sheep. AWGA records indicate he partnered with the Elorga brothers to purchase the Sancet and Fisher Sheep Outfit in 1922. In order to return to Spain to visit his parents, he sold his sheep to Alex and Joe Elorga in 1928.

Frank Sr. returned to Arizona and found work back in the sheep business with the Scott Sheep Company, which was headquartered in the east valley of the Salt River Valley. He would move on to another sheep company, and when that company went under, he and the other herders were given the sheep as their pay. As has been mentioned, many Basques were not paid each month, leaving their pay with the owner of the outfit. Since these herders and Frank Sr. had not been paid in four or five years, the sheep were turned over to them. Again, while not germane to the Elorga per se, many men left their monthly pay with the owner, and when it was time to settle up, the herder could be owed thousands of dollars. Outfits had to be sold to pay the wages, and even then, that money may not have paid the total sum owed.

In 1932, Frank Sr. began again with his own sheep outfit. He married Flora Locarnini and purchased an eight-acre spread across the road from Flora's dad's dairy operation.

Flora was one of six children that Joe Locarnini had. When Joe's wife died, he sent the children back to Switzerland to live with his sister. Four of the children would return, with Flora being fifteen in 1924 when she returned with her three sisters to help the dad run the dairy operation. They were responsible for milking forty cows each day.

Frank Sr. met Flora when he took a leg of lamb over to the family, and her dad suggested she do something nice for Frank. She did and married him! The couple had three children, Frank Jr., Rose Annlanolin, and Mary Alice. When Flora's father, Joe, died in 1936, Frank Sr. purchased the dairy farm from the sisters, and the couple lived in the adobe house Joe built. Its thick walls meant it was cool in the summer and warmer in the winter. The two- bedroom home served the couple well. Later, a new brick house would be built in front of this adobe home.

The Erramuzpes grazed their sheep that first summer on the Babbitt Winter Ranch located southeast of Winslow. The Wallace Allotment was in the Chevelon District of the Apache-Sitgreaves Forest. They shared the duplex with the Clarence Hancock family. The Erramuzpes had four rooms, one having a cement floor and the other three having wood floors. There was no electricity at the house, but they had the luxury of running water. The couple summered here for about ten years before they bought the Dutch Joe Ranch about fourteen miles away. The ranch was located equal distance between Winslow and Payson, fifty miles. A log cabin was located on the property that had been a ranger station. It had been built on the site as the Forest Service had marked each log so it could be assembled there. Flora said that the two-bedroom cabin had running water. "You went to the well and dipped for water, then ran to the house."[180] Butane lights were later added to the cabin.

Truck loaded with wool bags circa 1950s. Frank Jr. is at the top of the wool pile. *Photograph courtesy of Frank Jr. and Marcie Erramuzpe.*

Frank Sr. trailed his sheep from his property in Gilbert and then along the Heber-Reno Sheep Trail. In the summer after the sheep had started up the trail and school was out,

---

180 Doris French, comp. and ed. "Flora Locarnini Erramuzpe," *Arizona National Ranch Histories of Living Pioneer Stockman*, (2003), 115.

Flora and their three children would pack up and head to Dutch Joe Ranch. Packing meant taking the milk cow, chickens and chicken coop, washtub, an old gas-powered washing machine, and anything else they thought they would need for the trip. At this time, there was not a good road to the ranch, and they had an adventure just getting there. Once the chicken coop fell off their vehicle, and everyone scrambled to round up the chickens. When Frank Jr. graduated from Gilbert High School in 1952, his dad hoped he would go into business with him. He worked twenty-four years in the business, including the years he helped when he was a teenager.

The Erramuzpe Sheep Company had good years and bad years. The good years saw them paying off the property and having the ability to purchase more sheep. The bad years would try any man, but the family kept the business going through all the hardships. In one of the bad years, four hundred ewes were lost in a twenty-four-hour period. The sheep had just been brought down and put on winter fields. The maize milo had just been cut, but a few head of maize were left and the sheep ate it.[181] In another incident, half of the five hundred yearlings ewes that they had just purchased from Mike Echeverria were lost. The yearlings were trucked to where the other sheep were on the trail. There were two bands of the sheep of about one thousand six hundred to one thousand eight hundred per band, so the new ewes were split evenly between the bands. The sheep that

---

181 Corn, sorghum, milo maize, Sudan grass are toxic if they have been hit by frost or had a lack of water.

had been trailing had been eating the milkweed so they were used to it, but the new ewes were not and half of them died. In the fall of 1953, a hail and rainstorm killed 722 newly born baby lambs when a canal bank broke and drowned them.

Sheep being forced onto Blue Point Bridge over the Salt River circa 1950s. *Photography courtesy of Frank Jr. and Marcie Erramuzpe.*

Frank Jr. married Marcie Gomez in 1957. The Gomez family raised goats for mohair in the Safford area. The couple had five children that they raised in Gilbert. The couple told the author that Marcie raised their children in the log cabin that Frank Jr. had been raised in during the summer months. Marcie cooked using propane, and there was a gasoline-powered washing machine. One of the luxuries was a two-seater outhouse. Marcie was active in 4H, and she talked about the fun times with the kids she would take to the log cabin. The two-seater outhouse was popular with the kids as they liked to throw rocks at it when someone was inside. To protect the outhouse and cabin from forest fires, tin walls and roof were added. The rocks could make quite a sound hitting that outhouse and would scare anyone inside.

Cattle were introduced in 1961 to the Dutch Joe Ranch. Frank Jr. helped his father run the cattle and sheep business until 1969. That year, his father died after a long illness. All the sheep were sold to Felipe Perez of Glendale. Frank Jr. ran feeder sheep for about the next four years. He would buy sheep from Utah, Wyoming, the Navajo Nation, and the Zunis in Arizona and New Mexico. He would fatten the sheep and then sell them. Cattle would continue to run on the ranch until 1984. He would keep the cattle for five or six months and then send them on to feedlots.

The author asked Frank Jr. what made him decide to get out of the sheep business in 1969. He said, "Take a look around our property and all over the valley (Salt River), and all you see are houses clustered in subdivisions. We could not trail our sheep safely as people would not obey the laws that gave sheep on city streets the right-of-way. The subdivisions replaced their winter alfalfa fields. While other sheepmen moved further away to Casa Grande, I didn't want to uproot my children." The couple said that it had been a good life for them. They were able to instill in their children the Basque values of hard work and integrity. The children were required to help on the ranch during weekends and school holidays. Marcie and Frank are very proud of the accomplishments of their children and their contribution to the sheep industry.

Lockett Meadow near Flagstaff, Arizona where Arambel's grazed sheep in the summers. *Courtesy of Marty and Rebecca Arambel.*

Jean Arambel [182]

Jean Arambel arrived in the United States in 1937 from French Basque Country and made his way quickly to Arizona to be with his two sisters and their husbands who had arrived earlier. Jean had been sponsored by his sister's husband, Alejandro Elorga, who lived in Peoria. Jean tended his brother-in-law's sheep in the Buckeye area. He drove a

---

182 Personal interview with Marty and Rebecca, January 2016.

tractor for Emilio Cuevas as he built up his own sheep herd. Mr. Arambel built this while he worked for Ysi Otondo, another sheep owner. In the 1940s, Arambel was foreman for Ysi Otondo's outfit, the Frisco Mountain Sheep Company, who grazed sheep in the Buckeye Valley Area. Arambel would graze his nearly ten thousand sheep in the Mormon Lake area during the summer and in the Buckeye area on grazing leases during the winters. He trailed his sheep starting in Buckeye up by Lake Pleasant and on northward by the Black Canyon Trail, one of many trails used by the sheepherders. He partnered with another Basque, Pete Espil of Litchfield Park, in the 1960s and early 1970s. When Jean passed away, he was still in the sheep business. He discouraged his two sons from going into the sheep business.

Marty Arambel at summer cabin built by his father, Jean. Many sheep ranchers had similar cabins which were used for the family during the summer. *Photograph courtesy of Marty and Rebecca Arambel.*

The Etchamendys: Arnaud, Jean, JB, and Martin[183]

The first of the Etchamendys to migrate to the United States was Arnaud Etchamendy. He came in 1930. Sometime later, Arnaud became a citizen of the United States, which meant that he was now eligible to be drafted into the military during World War II. He sold his sheep to Frank Erramuzpe Sr. and headed off to war. When Arnaud returned from the war, he worked for Fermin Echeverria for a number of years. Arnaud began again in the sheep business with the help of Fermin and Fermin's son Felipe, allowing him to start the Diamond Sheep Company.

Jean, born in the French Basque Country in 1930, came from a long line of sheepherders. His father's family had been sheepherders in the Pyrenees mountains on the French side for countless generations. Jean was one of eight children. He realized at an early age that he would not be able to make his dream of having his own sheep company if he remained in France. While his father had sheep, he did not want to add to the flock as his sons did. Jean's father wanted his sons to run more cattle. Those years of watching and tending the family flock taught him many useful traits that would serve him well in the sheep business in the United States. With few opportunities available in his Basque homeland, he set out for greener pastures. At age nineteen, his uncle, Arnaud Etchamendy, agreed to bring him to the States and paid for his plane fare.

---

183 Personal interview with JB and Barbara Etchamendy, September 21, 2016.

Jean's first job with his uncle soon introduced him to a life as a sheepherder in Arizona. He was taken to the sheep camp in the area near Winslow. Jean was made the camp tender, a job packing the burros with the food that the herders would eat and cooking the meals for the other herders. He never really cared for cooking but he was fond of his job of handling the burros. Soon, he was given his own flock of two thousand ewes to care for after he had learned his uncle's way of doing things.

Until Jean had entered a French school in his homeland, his only language was Basque, but like Aleman and Espil, mentioned above, he taught himself English from a book for those who spoke Spanish, a language he also was unfamiliar with! He now could speak four languages: Basque, French, Spanish, and English.

Jean remained a sheepherder for his uncle for about eight years. During that time, he repaid his uncle for the airfare of bringing him to the States in 1949. As with many Basque herders, he took his pay in sheep and thus was able to begin his own sheep business, partnering with a fellow immigrant and friend, Jean Arriage. This partnership was the beginnings of the A and E Sheep Company.

The two Jeans ran their sheep from the winter grazing grounds around Parker to grazing grounds near Williams for two years. In 1959, two years after the partnership began, it was dissolved. That same year Jean Etchamendy married a Flagstaff native, Louisa Lopez. Shortly after his marriage, Jean became an American citizen. His wife, Louisa, was a teacher at Grand Canyon

Elementary School and was working on her master's degree at Arizona State University. The newlyweds spent their winters in the Scottsdale area, and Jean had his sheep winter on the alfalfa fields near Rural and Baseline Roads in what is now Mesa.

Upon the death of his uncle, Arnaud, in 1964, Jean bought his uncle's sheep and added them to his herd of the A & E Sheep Company. He formed the Diamond Sheep Company.

As late as 1976, Jean walked with his sheep to their summer grazing grounds in the White Mountains along the Heber-Reno/Morgan Sheep Trail and returned in the fall.[184] Jean walked the sheep from the pastures along the roads and desert to the beginning of the trail at Blue Point, the start of the Heber-Reno Trail. The sheep had the right-of-way on city streets and on the Blue Point bridge at the Salt River, the route used since the early part of the 1900s when it was designated as an official sheep driveway, but with more houses being built along the driveway, it became too much trouble to continue to trail the sheep. Trucking sheep to summer grazing grounds and then back to the valley for the winter was the only option left for many of the sheep ranchers, so Etchamendy began to truck his sheep. Jean retired in 1978 but was not happy being away from his first love, sheep. He bought himself a flock of two thousand sheep in 1980 and sold them in the mid-1980s. When he bought the sheep in 1980, he said, "I'll never leave the

---

184 Peterson, 2–9.

sheep business again. It is probable that I'll die with my boots on."[185] He could no longer find fields to graze his sheep during the winter months and for them to lamb. He moved to California.

Jean's wife, Louisa, was president of the Wool Grower's Auxiliary for many years. As a teacher of home economics, it made sense she would be a director of Make It Yourself with Wool for eighteen years.

After servicing in the French army in Africa in 1956, JB made the decision to migrate to the United States, just like his brother, Jean. In meeting with JB and his wife, Barbara, many stories were told, but the one that stands out is his trip to the States and meeting with family. In 1957, JB left his father and the rest of his siblings and headed to Paris to catch a plane to the United States. It just happened that at this time, the airline was on strike, and he told the author he was only willing to spend one more day in Paris. The next day the strike ended, and he was able to fly out. Like many immigrants and as his brother before him, JB did not speak English. He first arrived in New York and then took a plane to San Francisco. The stewardesses were concerned with his lack of English and followed him, unsure how this Basque was going to survive with no knowledge of English.

But he soon met a fellow Basque, Marcel Gastenaga. JB worked for his mother's cousin, Gracanne Labord, for the first year he was in the States. While he was there his brother, Jean and his uncle, Arnaud, came to visit him. He would

---

185   Paul Pollock, "Jean Etchamendy," *American Biographical Encyclopedia, Vol. 5* (Phoenix, Arizona: Paul W. Pollock, 1981), 254.

join them the next year. In 1965, he partnered with his brother, Martin, who emigrated after him, and they formed the Etchamendy Brothers Sheep Company. With a loan of $5,000, they were able to buy one thousand three hundred old ewes. The company saw a bumper crop in lambs born the next season and was able to pay back their loan by selling the lambs. The two brothers were able to purchase four hundred yearlings, and the company steadily grew. In 1966, they moved to New Mexico, where transportation costs were cheaper and grazing lands more plentiful. Their flock grew to two thousand ewes and bucks. By 1970, they had two thousand four hundred sheep. That year, Martin also went back to France, and JB ran the outfit alone.

JB married a woman he met at a cousin's party, Barbara. They had three sons, but he did not encourage them to go into the sheep business. He encouraged them to get an education as he did not see the years ahead getting better for the sheep business. There were many problems for any sheepherder. JB told of the problems of obtaining winter grazing pastures, the cost of these fields, and moving the sheep each spring to their summer pastures and then back to winter grazing. This became a challenge to him as well as other sheepherders who wintered their sheep in the Salt River Valley. Up until 1985, JB grazed his sheep on alfalfa fields in the Kyrene and Warner Roads area of Tempe. Many sheep could be found grazing these winter pastures, and JB told the author it was in the neighborhood of 125,000 sheep. Because of the pressures from people moving in and their demand for housing, subdivisions soon replaced the alfalfa fields. JB moved his operation south of Chandler in 1985 and grazed his sheep on the Salt River Indian Reservation. He continued to operate there until he sold out in 2000. His sheep were sold to an outfit in California, and they were moved shortly thereafter. As with many of the older generation of sheepherders, he was worried about the future. He told the *Ocotillo News*, November 17-30, 2001, "Someday we're going to be short of food. The farmers are getting broke, the world market is bringing everything from outside, and September 11 changed everything. I think this new generation should see that food doesn't come from the store."

While the three sons may not have gone into the sheep business, they would spend many hours with their father, helping him with the sheep. In the summers, the family would go to the summer grazing pastures and spend time together in fun and work. The boys earned spending money working for their dad on weekends and school holidays. They were paid an hourly wage. This continued during their college days too. Their parents paid for them to attend Arizona State University, an easy walk from their home.

JB, his wife, and two sons visited France in 1979 to see his dad and other family members. It had been more than twenty years since he had seen his dad. He took his sons to see the *echola*, the little one-room shack he called home during the summer grazing time of the sheep. He learned to make socks and sheep milk cheese while he was here every summer after being taught by his mother. When the author met the couple, they had fond memories of the six weeks the family had spent in JB's homeland.

"Echola" cabin in France where the sheepherder would spend his summer. These were permanent shelters unlike the tents used in the early days of the sheep industry in Arizona. *Photo courtesy of JB and Barbara Etchamendy.*

One of the interesting things learned about the Basque was that in their old country of Spain and France, it was not unusual for a Catholic priest to bless the sheep. Etchamendy, just like the Echeverrias, had the sheep blessed on a regular basis by a priest upon their return to the valley in the fall so there would be successful lambing.

The Etchamendys were active in the AWGA throughout their lives. Jean and JB were both members of the board of directors.

Father Charles Parker before blessing sheep with the family.
Photo courtesy of JB and Barbara Etchamendy.

The Danes

The Danes were another group that came to Arizona, with at least one family herding sheep. Niels Petersen, while not believed to be a sheepherder himself, brought many Danes from his home village who became sheep-herders, such as Hans Peder Thude. Hans purchased a 160-acre homestead and remained there until 1899, when he returned to Denmark and married. It is not known what Hans raised or grew on his homestead, but one of his sons, Gunnar, came to the United States in 1921 and stayed with the Petersons. He married one of their maids, and the couple settled down on property purchased near Price and Ray Roads. He began to run sheep in the mid-1940s. He purchased ranches in Heber, Williams, and Holbrook for summer grazing areas for his sheep. He ran the Paradise Sheep Company and Long Tom Sheep Company. He sold Long Tom Sheep Company to his daughter in 1968, and his nephew bought the Paradise Sheep Company in 1977. He is reported to have had as many as forty thousand sheep grazing in the East Valley around Chandler.

CHAPTER FOUR

NATIVE AMERICAN SHEEPHERDERS AND WEAVERS IN ARIZONA

The Navajo view sheep from a different perspective. While Anglos think of them as a commercial enterprise, the Navajo seldom sell the animal for money; the sheep means self-sufficiency for the family. They are family owned and provide food (meat, milk) and wool for both commercial and personal use. They may offer sheep for ceremonial use or sell to a friend; mostly they will trade.[186]

Through Navajo oral history and spiritual teachings, the tribal members believe that sheep have always been a part of their culture and lives.[187] Several Navajo told the author that they believed that something made the sheep disappear from their land, but with prayer, the sheep returned. Roy Kady explained the importance of sheep:

> Before we acquired the physical life of the sheep on this continent, we always held the idea of sheep in our genetic memory from thousands of years ago. We carried sheep fetishes in our pollen bags, and we sang the

---

186 Personal communication with Robbin Robinson, October 12, 2016.
187 Personal communication with several Navajo who wish to remain anonymous.

sheep lifeway songs for them. Our philosophy, spiritual-
ity, and sheep are intertwined like wool in the strongest
weaving. Sheep symbolize the good life, living in harmony,
and balance on the land. When we say "I will always walk
in beauty," that's what we are referencing.[188]

The story of the arrival of the Churro sheep into the
Southwest has already been described in a previous chapter. The Navajo-Churro descended from these first sheep.
Two Native American tribes that raided to obtain their
sheep were the Navajo and the Apaches. The Navajo most
likely obtained their sheep in the 1500s when they made
raids on the Spanish settlers in the Santa Fe and Albuquerque areas of New Mexico. Some scholars have suggested
that they may have obtained the sheep from raids in Sonora and Chihuahua, Mexico, but there is no substantial
evidence for this. The Apaches were not interested in the
sheep beyond wanting an easy food supply. Their part in
the story has already been told. The Navajo came to de-
pend on sheep for food and wool.

It has been suggested that the Hopi were given sheep
sometime between 1621 and 1630, when missions were established at Awatobi, Oraibi, and other pueblos.[189] As with
Father Kino, who supplied his missions with sheep as well
as other livestock and food crops, Franciscan missionaries
most likely also brought sheep and other Old World stock
and crops to those they witnessed to.

Not much else is known of the sheep bred by the Hopi

---

188 Cathy Short, "Lamb to Loom," *First American Art* 9 (Winter 2015/16), 35.
189 Towne and Wentworth, 35, 39.

or Navajo until the 1850s, after the US military made excursions across the northern portion of the state. From Lieutenant Ives, more known for his exploration of the Colorado River by steamboat, it is known that the Hopi had flocks of black and brown sheep in 1851.

While most Navajo were devoted to herding their sheep, which required little work on their part, the younger ones were still warriors. General Carleton reported to Washington that a number of citizens were killed and livestock was stolen by these warriors. In 1850, a military inspector estimated that the Navajo stole at least 47,300 sheep in eighteen months. While the exact number of stolen sheep will never be accurately known, the raiders did not take all the sheep, as "they prefer leaving a few behind for breeding purposes in order that their Mexican shepherds may raise them new supplies."[190]

James[191] stated that the inferior nature of the Navajo wool lent itself to making excellent rugs and blankets. It had been stated, though, that by American standards the Navajo were not good stockmen. James further wrote that scientific breeding and care of the sheep was unknown to them. James[192] quoted Dr. Letherman as saying that males or rams were kept with the females all year, and thus lambs could be born at any time of the year, resulting in considerable losses especially if the lambs were born during the winter months. These newborns increased herd size, but

190 Josiah Gregg, *Commerce of the Prairies* (Philadelphia, PA: Lippincott, 1962), 123.
191 George Wharton James, *Indian Blankets and Their Makers*, (Glorieta, NM: Rio Grande Press, 1914), 56.
192 Ibid., 208.

the increase would be slow with this practice. Even by the 1860s, the Navajo were reported to have hundreds of thousands of sheep grazing on their tribal lands.

The raids were becoming very tiresome and more frequent, so the federal government declared war on the Navajo in 1863. Kit Carson, along with a regiment of volunteers, was so successful as he swept into Canyon de Chelly, killing all the livestock and destroying fruit trees and crops, that by 1864, nine thousand of the Navajo had surrendered and were forced on the Long Walk to Bosque Redondo, New Mexico. A few Navajo avoided capture and remained hidden in remote canyons with their sheep. The Bosque Redondo internment area did not appeal to the Navajo, and they spent months appealing to the War Department to be allowed to return to their old tribal grounds. During the years of fighting and living in New Mexico, their herds of goats, sheep, and horses had been drastically reduced. When approved to return to their homeland in 1868, they were allocated land, farm equipment, seeds, provisions, a thousand goats, and fourteen thousand sheep![193]

The federal government encouraged the Navajo to increase their flocks and distributed Merino and Rambouillet breeds. These two breeds were "considered better suited to the East Coast lamb and wool markets."[194] These breeds and others introduced helped to dilute the Navajo-Churro sheep.

---

193 Wentworth, 243.
194 Short, 37.

The Navajo-Churro sheep has long legs, a narrow body, and light bone structure. There may be as many as four horns on both ewes and rams. They have a "double-coated fleece that weighs four to six pounds. The fine, soft inner coat provides insulation, and the long, coarse outer coat protects the inner coat from dust and dirt while repelling rain and snow."[195] The sheep is known for its ability to easily breed. Males can weigh an average of 160 to 200 pounds and an ewe, 100 to 120 pounds.

An 1862 government report gives some insight into the Navajo flocks. It stated that the flocks vary in size from three hundred to four thousand ewes, wethers, and bucks, all run together all year. Lambs occur all year. It further stated that there was little desire to improve their flocks "and when good bucks are given them by the Indian agents, they are very apt to barter them for other stock."[196] The wool is coarse and used for carpets and blankets. The mutton is used on the reservation.

As sheep increased in numbers on the reservation, the federal government began to require a reduction in the herds. It was believed in the 1930s that the reservation was being overgrazed. The atrocities of the stock reduction program during this time were horrific in many ways for the tribe. There was a demand for reduction of fifty thousand sheep and one hundred and forty-eight thousand goats. The federal agents shot many of the sheep and

---

195 "Navajo-Churro Sheep", *The Livestock Conservancy. Heritage Breeds,* accessed October 1, 2016, www. thelivestockconservancy.org.

196 U.S. Department of Agriculture, Bureau of Animal Industry, 1892, 945.

goats when the reduction program was not moving as fast as they thought it should. The shot animals were left to rot in the ditches where they had been pushed. Some of the dead animals were partially burned. In many instances, this was done in front of grieving family members. The owners of the livestock, the women, openly criticized the actions of the government and its agents. This wanton destruction of the goats was especially hard on families that relied on these animals for milk, cheese, and meat.[197] Non-Navajo believed goats had little market value. Again, the federal government failed to understand the subsistence economy, where goats were important as a dependable source of food. Also, in reducing stock, there was a failure on the part of the Bureau of Indian Affairs (BIA) to "understand Navajo concepts of stock ownership." The men in the families were credited with ownership instead of the women. In 1936, the Navajo women rebelled against the federal government.

> In general, the BIA tended to ignore long-established cultural patterns regarding livestock management, and they often disparaged local knowledge and cultural understandings of the environment. about implementing the livestock reduction program, they refused to solicit or listen to Navajo advice. Finally, BIA officials tended to be sexist in that they disregarded the role of women in Navajo society.[198]

---

197 Short, 37
198 "The Navajo, Sheep, and the Federal Government", *Native American Netroots*, November 10, 2011, assessed October 1, 2016, www.nativeamericannetroots.net/diary/1136,

Navajo-Churro sheep with four horns.
Photograph courtesy of Robbin Robinson.

In 1934, the US Department of Agriculture established the Southwestern Sheep Breeding Laboratory at Ft. Wingate, New Mexico. Its purpose was to determine the breed of sheep that would thrive in the type of vegetation and climatic conditions of the region or the high desert. After thirty years of looking at "original old type," they concluded that "old type" would be best suited for this terrain and its wool would be the best type for weavers.[199]

In 1977, the Navajo-Churro sheep had decreased to approximately four hundred and fifty head. Not wanting further loss of the breed, Dr. Lyle McNeal formed the Navajo Sheep Project (NSP). "Its primary purpose was to locate, identify, rescue, and begin a scientific genetic process of saving the original Navajo-Churro sheep from extinction."[200] The first Churros were donated from a ranch in California. The donor had traced his sheep back to the Southwestern Sheep Breeding Laboratory. In the next two decades, other Churro breed sheep were found on the Navajo Reservation from families living in the most isolated places. Over the next twenty-five years, the location of the NSP changed, but the focus never changed as individuals within the project kept trying to replenish the stock. Wool from the NSP was first used by the famous Burnham, New Mexico, Navajo weavers for their textiles in 1984. The NSP experimented with food products such as lamb meat and pelts to increase cash

---

199 "Preservation An Endangered Breed", *Navajo-Churro Sheep Association,* assessed October 1, 2016, www.navajo-churrosheep.com/index.html.
200 "The History of the Navajo Sheep Project Organization", *Navajo Sheep Project,* assessed October 1, 2016, www.navajosheepproject.com/nsphistory.html.

flow in 1987. Churro summer sausage was marketed in 1989, and they experimented with milk, mutton stew-stick, and Kosher lamb jerky in 1992 and 1993. Breeding sheep were introduced to various weaving groups. To increase the survival rate of the breed from predators, llamas were introduced and reduced losses. The Churro was sent to Chihuahua, Mexico, to introduce a new gene pool. A new name that "reflects accurately the efforts and programs of the NSP, which is the Navajo Sheep Project: Serving People, Preserving Cultures" is voted on and approved by the NSP Board of Trustees. [201] Other problems occurred during these years with NSP, but the project survived. The history of the NSP and its contribution to the livelihood of the Navajo has been well documented.[202]

Navajo-Churro sheep and other breeds grazing on the Navajo Indian Reservation. *Photograph courtesy of the author.*

---

201 Ibid.
202 Ibid.

Egan writing for the *Salt Lake Tribune*, Utah on May 24, 1998, discussed that with the Churros, a better quality of wool was possible, and thus livelihoods increased. The wool from a Churro sheep is cleaner with less grease and longer, straighter fibers. This makes stronger thread when spun. It contains little lanolin. The women still spin wool into thread and weave it into rugs and blankets. These rugs and blankets have been highly prized ever since Paul Harvey introduced this art form to the tourists he brought to the Southwest through his hotels and their gift shops along the path of the Santa Fe Railroad.

In 2012, the Navajo Nation had 9,378 farms raising sheep and lambs with a total number of 171,107 sheep and lambs. The reservation has five main areas, and each are divided into districts that the US Department of Agriculture (USDA) uses to record sheep and lambs on the reservation.

Henry Lane, age ninety-eight, told a reporter in 2011 that the young people did not want to herd sheep as was tradition on the reservation. Lane had been herding sheep for about eighty years off and on and had bought a modern convenience to help him: a truck. Avery Denny, from Diné College, told the reporter that traditional methods have changed. He said that in the 1960s, walking or riding horses with the sheep grazing in front was their way, but then vehicles came into use, just as Lane now uses. "Denny said he sees fewer Navajo youth embracing the sheepherding tradition and blames the waning enthusiasm on a quest for higher education, good-paying jobs, and the conveniences of modern life."[203] Lane had only one child, a daughter, who was interested in keeping the family tradition of sheepherding. Lane was proud of his stock of sheep as they were "descendants of his late aunt's herd from 1913."[204]

A journalist, Michael Benanav, on July 27, 2012 with the *New York Times* described his experience of following the "traditional contemporary life of Navajo pastoralists." Benanav described how Irene Bennalley trailed her sheep and goats to the Chuska Mountains that summer. This seasonal animal migration had been a way of life for many tribal people for centuries. The journalist further stated that few still undertake this migration, and those that do are the women who "weave the rugs for which the region is famous." As the author of the article discovered, her home for the months of June to October had no electricity and no running water. Whether is it still true that some tribal members on the reservation are beginning to allow more cultural tourism, Irene shared her life with the journalist for three days. The journalist experienced what few who do not live on the reservation had experienced, a glimpse into their traditional ways. "The sheep are like our parents," Irene told Benanav. "They feed us and give us comfort from the cold."

One group fighting to preserve the Navajo life is Diné bé iina, a grassroots nonprofit organization founded in

---

203 Betty Reid, "Arizona's American Indians – Sheepherder", assessed October 1, 2016, archive.azcentral.com/news/native-americans/?content=sheepherder.
204 Ibid.

1991. The word itself means "way that the people live." It promotes sustainable livelihood, which is sheep, wool, and weaving. It is "dedicated (to) restoring the balance between Navajo culture, life, and land." For the past twenty years, there has been an Annual Sheep is Life celebration held in June at Diné College in Tsaile. There are educational workshops, Navajo cuisine, and the Navajo-Churro sheep.

Carpenter, William J., Copyright Claimant. Home Industry. ca. 1915. *Image. Retrieved from the Library of Congress, https://www.loc.gov/item/2005686549/. (Accessed October 22, 2016.)*

For the past sixteen years, Robbin Robinson has been raising Navajo-Churro sheep on her ranch in the Flagstaff area. She was always around animals and was a Future Farmers of America (FFA) member, and she wanted to raise animals that were indigenous to the Arizona landscape, thus the Navajo-Churro sheep. The sheep spend the summer out on pastureland. In winter, the flock is brought back to her ranch. She trades her animals to families on the reservation, which helps keep the genetics fresh. Robbin shears her sheep herself and gives the wool to Navajo friends or to those who would like to learn to weave but do not have any wool of their own. She is an inspector for the Navajo-Churro (NC) Sheep Association. In addition to inspecting mail-in registration applications from around the country, an inspector is called to ranches to determine if a Churro sheep shows all the characteristics of the Navajo-Churro breed and can be registered with the association. Because the Navajo started mixing the government-issued rams of other breeds with their Churro sheep many years ago, the association works to ensure today's animals do not exhibit characteristics of those other breeds.

Robbin also helps coordinate the Flag Wool and Fiber Festival held the first weekend after Memorial Day weekend. The purpose of the festival is to encourage appreciation of ancient materials and weaving techniques. The festival has celebrated twenty years as of 2016. Robbin and the other coordinators changed the emphasis of the festival to include shearing, which fills a niche for families with small herds that cannot afford to hire a shearer. Some of the animals had not been sheared in years, she told the author. These families raise the goats, sheep, alpaca, and llamas for the meat that some sell. When they had their animals sheared, most gave their wool to weavers and a few sold the fiber. Robbin said that the festival purpose is to educate the public and the animal owner. The public get to see how textiles are created from growing on the animal to the shearing, wool preparation, dying, yarn spinning, and finally the textile. Animal owners learn proper care of the animal to maximize their fiber output. Different animal breeds were on hand to educate the public about each one.

Navajo-Churro ewe with lambs.
Photograph courtesy of Robbin Robinson.

In an interview in the *Arizona Daily Star*, March 31, 2006, it is believed that by raising her Navajo-Churro sheep and being actively involved in educating the public about wool through the Flag Wool and Fiber Festival, "she (Robinson) is helping to preserve a true Southwest treasure."

CHAPTER FIVE

THE YEARLY CYCLE OF SHEEP RANCHING

It's amazing that the sheep in Arizona even existed let alone flourished given the amount of adversity in the beginning, including Indian raids, drought, and range wars. Yet the sheep industry managed to grow and continued to make huge strides, according to the reports on the sheep in the early part of the 1900s. The reports tell us about the health of the industry, describe the land the sheepmen used for grazing, and provide an account of the yearly cycle that the ranchers went through each year. The diagram below shows this cycle for the last three outfits operating in the state.

Trailing

Mark Pedersen told the Mesa *Tribune*, October 6, 2007 "he gets a kick out of watching people in the East Valley stop and watch as the sheep move past buildings along paved streets". Pedersen has even seen spectators set up lawn chairs to watch the animals walk through town. "It is fascinating for some people," he said. "A little bit of history that a lot of people in the hustle and bustle world of the big towns get to see." People may not know that the movement is important for the overall health of the sheep and improves lambing in November.

The Yearly Cycle of Sheep Ranching.
Drawing courtesy of the Pinal Ways. Summer 2010.

In the past, the sheepherders trailed their sheep during the spring and fall between the north and central portion of the state. In the late spring, usually late May, they would trail the sheep northward to the summer grazing fields across what are known as "driveways." These driveways were established by the Bureau of Land

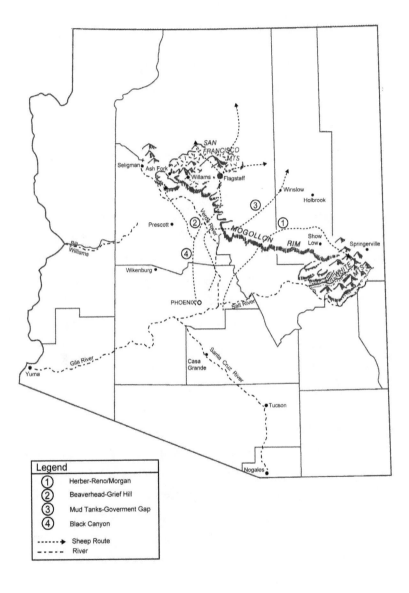

Sheep Driveways. *Map drawn by Francisco J. Caraveo.*

Management and the National Park Service. The driveways protected the natural vegetation and native animals. In 2011, the Heber-Reno/Morgan Mountain Sheep Trail was used for the last time, leaving only the Black Canyon and Beaverhead-Grief Hill Driveway. Both the Manterolas and Auzas continue to use a portion of this historic sheep driveway. Their sheep are trucked to Bloody Basin and then trailed the rest of the way northward along the last remaining driveway.

Originally, the driveways belonged to the AWGA. Sheep owners paid dues to the association, and they bought the Black Canyon and Beaverhead-Grief Driveway, which ran between Black Canyon and Flagstaff. Four sections totaling 2,560 acres were purchased at seventy-five cents per acre.[205] The Livestock Homestead Act of 1916 recognized this trail and others used by the livestock and regulated the width and permits that controlled the movement of the sheep across forests. Boundaries minimized disruption of cattle in the areas. The Taylor Grazing Act of 1934 regulated the use of forestlands under the national forests, and then control was given to the Bureau of Land Management. The driveways are legally described in Executive Order No. 2464-A, which was signed by President Woodrow Wilson. Several sheep ranchers told the author that in the 1970s a challenge was made about an association owning deeded land, and the judgment was that the land

---

205 Arizona Wool Growers Association, NAU.MS.233, Cline Library, Special Collections and Archives Dept.

had to be sold with all profits distributed to members of the AWGA.

The Dobsons would start from their winter grazing at Baseline and Signal Butte and go through town and pick up the Heber-Reno/Morgan Mountain Sheep Driveway at Brown and Signal Butte. When the sheep were trailed along city streets, they had the right-of-way. Police escorts were sometimes used, depending on the situation.

Police escorting sheep through Mesa.
Photograph courtesy of Cindy Shanks.

Dwayne Dobson stated that the trail has stayed pretty much the same since his grandfather's day, except that it is longer with all the development in the valley that they have to skirt the animals around. Rapid population growth along the route used to get to the trails has decreased the ability to trail the sheep between winter grazing areas and to/from the northern summer grazing pastures. There is also private land along the trail

The Original Verde River Sheep Bridge, at the time called the Red Point Sheep Bridge, with sheep crossing about 1944. (Tonto National Forest), Cave Creek, Maricopa County, AZ *(Courtesy of HAER ARIZ,13-CACR.V,1—29)*

that needs to be skirted. In the area of Silver Creek south of Holbrook, Dwayne said landowners landscape with rocks and bushes, and when the sheep come through, landscaping and sheep don't mix well.[206]

The trails crossed rivers, and bridges were needed to facilitate crossing the sometimes raging waters. Many sheep drowned before the bridges were built. The Verde River was a formidable obstacle, with its swift current. The sheepherders could expect to lose a few sheep every time the river was crossed. Thus, there was a need for the bridges. At first, temporary bridges were built like the "pontoon-type bridge at Red Creek (six miles north of the Verde) and a small suspension bridge at Tangle Creek." They were used for about three years and were reliable when the water level was low in the river. The problem was that the bridge had to be dismantled after use to keep it from washing away during high-water levels.

When Frank Auza was foreman for the Flagstaff Sheep Company, he built a permanent bridge in 1943 with the help of a local builder, George F. Smith. Dr. Raymond was the owner of the Flagstaff Sheep Company and partnered with Ramon Aso in the Howard Sheep Company. There were three grazing allotments in the area, and two were used by these companies. Dr. Raymond made an application to the Tonto National Forest for permission to build the bridge on one of his allotments. Permission

---

206 Personal communication with Dwayne Dobson and Frank Erramuzpe Jr., fall 2016.

was received in March 1943. The bridge was designed by Mr. Cyril O. Gilliam of Phoenix. Originally, it was called the Red Point Sheep Bridge, but today it is known as the Verde River Sheep Bridge. The bridge was a cable-stayed suspension bridge and was "erected with hand tools and a few mules."[207] Its length along the walkway was 476 feet, and the bridge generally ran east to west. Some of the materials for its construction were salvaged from the old Blue Bell Copper Mine near Mayer. The huge supporting suspension cables came from that mine, and Mr. Auza and Harry Cordes trucked those materials to the bridge. The guy cables came from the Golden Turkey Mine. These two mines were located in the Bradshaw Mountains, southwest of Mayer, Arizona, but it wasn't simple to build the bridge. Other materials, such as the wood for the planks had to be secured, and the only way supplies could be procured was to get permission from the War Production Board.

At one time, there were questions as to when the design plans were made for the bridge. In a report prepared by Gerald A. Doyle & Associates, historical architects, the design for the bridge "was prepared in time to have the drawings referenced in the Flagstaff Sheep Company's application to the War Production Board for authorization to obtain construction material."[208] A letter, dated March 23, 1943, from the War Production Board, stated what ma-

---

207 Information on kiosk at bridge. Most of the description about the bridge comes from the kiosk.

208 Gerald A. Doyle & Associates, P.C. "Verde River Sheep Bridge" (Phoenix, AZ: July 1987), 7.

terials could be purchased, and an amount that could be spent was sent to Ramon Aso.[209]

Sheep frolicking through a highway underpass. It was also used as a counting point. *Photograph courtesy of Cindy Shanks.*

The logistics of getting all the material to the site of the bridge was also a challenge as the roads were dirt roads, and only small sections coming from Cave Creek are paved today. The bridge took less than six months to build, being built March through June. Less than a year had passed when it was realized the support towers needed to be concrete instead of wood. A kiosk at the bridge states

---

209 Letter with other documents found in Howard Sheep Company folder, Arizona Wool Growers Association, Cline Library, NAU.MS.233 Series 1.2.40

"the eleven thousand sheep that wintered on these three nearby Forest Service grazing allotments could safely cross from one side of the Verde to the other." The bridge was used for the first time in the fall of 1943 when Frank Auza was the first to cross it on his horse.

Joe Manterola paid Dr. Raymond for the bridge in September 1945 and received money from others who would use the bridge. "A special permit is issued to Jose and Marianne Manterola, Jose Echenique, and Mud Lake Livestock Company to use one acre of national forest land for maintaining a sheep bridge across the Verde River."[210] The bridge allowed the four sheep companies—Flagstaff Sheep Company, Manterola Sheep Company, Mud Lake Livestock Company, and Howard Sheep Company—to safely move their sheep across the river. Concrete slabs could be seen at the west side of the river, evidence of a long-ago ranch headquarters with a ranch house, shearing shed, and corrals. Wool from the sheared sheep was hauled to warehouses in Phoenix to the wool buyers.

According to the kiosk at the bridge,

> Although several other bridges of this type previously had been built in Arizona, Verde River Sheep Bridge was the last of its kind in the Southwest when it was entered into the National Register of Historic Places in 1978 (Actual date was November 21). In 1988, weakened by years of service and floods, the bridge was disassembled. In 1989, the bridge found here today was erected. Reminiscent of the original structure, the new bridge calls the ingenuity of the pioneer sheepmen and a way of ranching that has virtually disappeared from the Arizona scene.

---

210 Gerald A. Doyle & Associates, P.C, 24.

Historic photograph, photographer unknown, c. 1954. VIEW OF LOADING SACKS OF FLEECES ONTO TRUCK AT JUNCTION OF SEVEN SPRINGS ROAD AND TANGLE CREEK ROAD. - Verde River Sheep Bridge, Spanning Verde River (Tonto National Forest), Cave Creek, Maricopa County, AZ HAER ARIZ,13-CACR.V,1—41.

Sheep Bridge over the Verde River north of Cave Creek
built in 1989. *Picture courtesy of the author.*

A family tradition began in 1883 lives on at Cordes. John and Lizzie Cordes bought a small adobe stage stop and saloon that became a gathering place for sheepmen on the Black Canyon and Beaverhead-Grief Hill Trail. An application for a post office was approved, and John became the first postmaster. They began to offer many of the same services that James Houck offered at Cave Creek to attract the attention of passing sheepmen: a blacksmith shop, general store, stables, a boarding house, and dipping vats. The author was told at least 125,000 sheep were dipped in the early 1940s and 1950s. A shearing plant also operated here. It was a little mecca in the middle of nowhere, a place where seven driveways passed. A spur railroad, the Bradshaw Mountain Railroad, was built in 1902, and a small siding with a stockyard and a warehouse was added to the couple's business.

John and Lizzie's grandson Henry bought the store from his father in 1937. He continued to operate the store and branched out into other businesses to diversify his chances of survival in bad economic times. He helped Frank Auza get some of the materials from the Blue Bell Copper Mine for the Verde River Bridge. While not germane to the history of sheep, he continued keeping precipitation records, just like his father, and then added temperature. The store remained open and supplied the sheepmen until the late 1960s.

His granddaughter, Cathy, reopened the store in 2005. In May 2010, the last two families continuing the traditional trailing, the Manterolas and the Auzas, brought their sheep through the area after trucking them to Bloody Basin. John and Lizzie's great-great-granddaughter, Cathy, hosted a barbecue at the historic station to celebrate their arrival. The store is full of memorabilia, with antiques and pictures of bygone days. She operates the store on certain weekends each year. Cathy was unable to speak with the author and may have shed more light on this ranch and its role in sheep trailing.

Sheep with shepherd and dogs.
Photography courtesy of David W. Schafer Photography.

Back on the trail, usually only two herders would accompany a band of about two thousand sheep. The herder knew the trail, and the sheep who had been on the trail before knew the route. Rocks can be found in certain locations, marking the trail. The herder's job was to watch the herds for signs of injury, which often occurred as the sheep's feet and mouths became sore from embedded cactus thorns. The herder's day begins at daylight after a hearty breakfast, and he starts the day guiding the sheep. At about 11:00 a.m., the camp tender unloads the burros and prepares lunch for the herder. The herder and sheep will rest for a couple of hours, and around two in the afternoon, the sheep will start walking and continue grazing along the way until sunset. The sheep will cross streams and under freeways along the driveway, following the same route every year. The sheepherder was an important part of the operation.

Rocks marking Black Canyon Driveway at Verde River crossing. A few of these are still found on the trails today. *Photograph courtesy of the author.*

The camp tender would take care of the supplies, pack the donkeys every time it was necessary to move, make camp, and be the cook. The donkeys carried water for use in the camps, a Dutch oven for cooking their food, food supplies for both men and the dogs, at least one rifle, the personal supplies of the herder and camp tender, their tent and bed rolls, and many other useful items. The camp tender would move ahead of the flock and prepare the camp, cook the meals, and erect the tent for both of them. The Great Pyrenees dogs were always nearby in case of danger from predators.

The sheep were moved from one grazing area to another, depending on the requirements of the grazing permit. This way of herding has been done for many years and was a part of the herding tradition of the Arizona sheep industry and ranchers.

The author joined three generations of Auzas and family friends as they came to watch and help the sheep cross the Verde River at Thousand Peaks near Camp Verde. This event used to occur regularly with many sheepherding families in the state but now is limited to two families. It was a very smooth operation and was over much faster than the author had wanted it to be. The sheep had been mostly lying down prior to the herders packing the burros with crates and food for the days ahead.

Two types of dogs, the stock dog and the security dog, traveled with the flock, and both were equally important. The stock dogs manage the flock, getting the sheep moving down the road or out on the range to fresh pastures. These dogs are a collie breed. The Great Pyrenees was the security dogs protecting the sheep from predators, whether four-legged or two. It is ironic that one of the sheep's predators is another canine, the coyote. Packs of wild dogs also can cause much destruction in a herd of sheep especially when the ewes have just given birth.

Siesta hour before crossing the Verde River. Every day
on the trail the sheep and herders will take a break
from the heat. *Photographs courtesy of author.*

Auza's sheep crossing Verde River May 2016. Men are downstream to protect sheep from being carried away by current. *Photograph courtesy of author.*

The time for the sheep to be trailed prior to reaching their summer grazing allotment depends on which driveway was used. In the past, the animals could be on the trail a minimum of forty-five days, being driven three or four miles per day. June 1 is the first day that sheep can enter the forest grazing allotment. The sheep remain on the different allotments assigned to each sheep rancher for a specific number of days determined by the Forest Service in that area. Checks are made to ensure the sheep are moved after the time period has expired. Each sheep rancher pays a specific amount per month per sheep to use the allotment. Many sheepmen interviewed told the author that it was cheaper to pay grazing fees than the taxes on deeded land.

Joe Manterola with Great Pyrenees. *Photograph courtesy of the author.*

While on the trail, the Basque sheepherders and the children of the sheep ranchers engaged in the art of arborglyphs or tree carvings. The carvings can be found in aspens, birch, and beech trees. Joxe Mallea-Olatxe referred to this art as "a living outdoor museum that stretches from Washington (state) to Texas and from California to North Dakota."[211] Mallea-Olatxe said that the carvings are

---

211 Joxe Mallea-Olaetxe, "Carving Out History: The Basque Aspens," *Forest History Today* (Spring/Fall 2001), 46.

considered both folk art and recorded history as they document the Basque way of life. It has been determined that this art has been almost exclusively done by the Basque since 1850s to 1970s. Unfortunately, the trees have a limited lifespan because of decay, vandalism, or fire.

Sheepherders could enjoy creating the artwork, and other sheepherders could view it. Mallea-Olaetxe has categorized them into four major topics: personal information (name, birthplace, region), sheepherding (location of water holes and directional information), loneliness, and erotic fantasies.

Trailing was not without its hazards, however. Poisonous weeds like locoweed, predatory animals, and the wandering ewe or even many wandering ewes could ultimately reduce the bottom line for the sheep owner. Coyotes were a big threat to the herd, and when the Mexican gray wolf was reintroduced to the eastern portion of the state in the 1990s, herders and their dogs had to be constantly on the lookout. The herder was also responsible for the health of the sheep. He needed to be able to spot unhealthy ewes and treat the animal accordingly.

Lightning could also kill the sheep, herder, or dog. One outfit lost thirty sheep in one storm. The animals were under cedar trees, and the lightning traveled down the trees. Mr. Erramuzpe told the author that a vacuum is created and literally pulls the wool off the animal. He and his wife remember seeing the wool in the tree branches for years afterward. Another incident killed approximately one hundred and twenty sheep. One outfit lost both a

Aspen tree carvings. Some can be very simplistic while others are more detailed. *Photographs courtesy of the author and Cindy Shanks.*

sheepherder and his dog. A sheepdog was killed when he lay against the metal pole holding the tent, and the lightning traveled down the pole.

Shepherd and his dog with his sheep.
Photograph courtesy of the author.

Today, the herder and camp tender use cell phones to order supplies. When supplies are brought to them, they also receive any mail from home. The manager, who speaks English, will interpret for them and makes sure that they get medical attention whenever it is needed. The manager also settles the many petty disputes that may arise from the two men living in close quarters. In the past, the camp tender would travel many miles with the burros to get his supplies from the ranch, which had already purchased them in town. The camp tender collected mail that may have accumulated. He would load the burros and either that day or the next would return to the camp. He would gather any interesting news of what was happening to pass to the herder.

After the sheep reach their summer grazing land, the rams are trucked northward and put with the ewes in mid- to late June. The ewes are run in a close flock to allow the rams to ferret out those whose ovulation period has begun. It is usually one ram to fifty ewes. Toward the end of the summer grazing period, the ewes are tagged, which involves shearing the area around their eyes to prevent wool blindness and shaving their stomachs and rump areas.

Tagging is an interesting enterprise. The shearing crew and herders build temporary corrals and a shearing shed. The dogs help the herders get the sheep into the corrals the night before. The day begins with the sheep run along a chute toward the shearing shed. They are pulled into the shearing shed, and wool is sheared around their eyes, stomachs, and rumps. The last two areas are sheared to help newborn lambs to milk and for their birth. They are then released out to a holding area, where they are checked for any infections and given an injection for over-eating. Counting is the last step as this tells the owner his decrease in numbers from lost ewes and rams minus what the camp tender and herder are allowed to consume. Next, the rams are sheared, given an injection, and branded. They are then released into a holding pen.

Water hauled to sheep either while on the trail or once in their summer grazing area. Water holes were created by the rancher but if the rains did not supply the needs of the sheep, it would be hauled in every few days. *Photograph courtesy of Liz Espil Mooney.*

The wool and fleece are collected and bagged by one man.[212] His job is to help when ewes or rams jump the corrals or are uncooperative. A machine is used to bag the wool today instead of it just being shoved into bags. The

---

212 Wool from one sheep is called a *fleece,* whereas *clip* is used for many fleece.

rancher will keep the wool until all the sheep are sheared in the spring. Not all outfits operated the same way. Some outfits completed shearing the animal prior to lambing.

Tagging (removing wool from rump and around eyes) of the sheep. *Photograph courtesy of the author.*

For another couple of weeks, the sheep will remain on the summer grazing allotment before they are trucked south to their winter grazing. The forest grazing allotment permit will end by the end of October. Most of the rams are separated after tagging. Some of them will remain with the ewes that may have not been impregnated before. When it is time to truck the ewes southward, the pregnant ewes are handled carefully as they are loaded on the trucks. The ewes go in a corral that has been built, and the pregnant ewes are loaded first on the trucks. More space is required for each ewe. No one wants premature births. With limited trucks, many trips are needed. Commercial livestock trucks may be hired if a family does not own their own trucks. Experienced drivers with commercial driver's licenses are needed to ensure the safety of the ewes and rams.

Bagging of the wool after the tagging process.
Photograph courtesy of the author.

Sheep with wool in their faces and after tagging. Wool not removed from around the eyes can led to wool blindness. *Photographs courtesy of Elizabeth Espil Mooney and author.*

Sheep being unloaded in winter grazing area.
Photograph courtesy of Carmen Auza.

Since not all sheep are moved at the same time, some of the herders and dogs will remain with the sheep until all are loaded on the trucks and moved southward. The herder who remains with the animals keeps what he needs, but all other equipment and supplies are taken south. The manager's responsibility is to accompany each load every day, and this may take several days. Loading is done early in the morning, and the manager with the livestock trucks head southward. The ewes are unloaded in the late afternoon, and then the process begins again the next day. The manager will usually drive back that night to begin the process all over again until all the sheep and herders have been moved to winter grazing.

Once the sheep arrive in their winter grazing pastures, it is a very intense time for everyone at the ranch. Something always needs to be done so extra employees are needed. Portable electric and page wire fences are built and constantly torn down and reset as the sheep move from one pasture to another. Sheep are often walked between fields down the roads. As the ewes get heavy with the lamb or lambs she is carrying in the last gestation period, she is fed supplemental feed. Lambs begin to be born November into December. The day begins at dawn and does not end until sunset if there are no baby lambs to feed or other problems during birthing.

As previously stated, some lambs have to be bottled-fed, a full-time occupation requiring regular feeding times. Many children told the author that this would be their job. The author was told that a good sheepman can tell the history of breeding and lamb production for each and every lamb, even though there may be upward of two thousand lambs in his care.

Author and herder bottle feeding a leppie. Leppies can be a twin the ewe rejects or a triplet which she cannot feed as this little one was. Triplets are a rarity, but one year the Erramuzpe's had 80 triplets. *Photograph courtesy of Carmen Auza.*

A ewe that gives birth to twins is kept as she will always produce twins. An outfit that can double the number of animals because of twin births thereby makes a large profit margin. The two twins are hobbled together to give the weaker one a chance to survive. Some ewes reject their babies, so these must be bottled feed. A ewe that losses her lamb will be paired with an abandoned lamb so that it may survive. These ewes are separated and watched sometimes for a night to ensure they do not reject the orphaned lamb.

Two other procedures are also undertaken: docking and castration. Docking is the process of shortening the tail. The overall health is improved as docking prevents fecal matter from accumulating on the tail and rump of the sheep. Docking can facilitate shearing as the tail is not in the way of the shearer. Only a certain number of adult sheep are required for breeding purposes; all other male lambs will be castrated by using a very small rubber band about a quarter-inch wide that is stretched by a special tool. The band is placed at the base of the testicles. The tight band closes off the blood flow, and the testicles die and fall off naturally.[213]

In years of good winter rains, the sheep will be moved out to the desert to feed. The ewes and the lambs do better when fed on desert vegetation. This helps reduces the cost to the owner as he is not paying to lease alfalfa fields. Every means to reduce cost is undertaken as the profit margin is low today with the hiring of foreign herders, grazing fees, and leasing fees, just to name three costs. Many sheep owners told the author that in the past, the price they received for wool would cover their yearly expenses, and the lambs sold were their profit.

The lambs are shipped to market in late March or early April; these are called the spring lambs or Easter lambs. The process begins again.

Whenever shearing takes place, either prior to the birth or afterward, professional shearers are hired from

---

213 In the past, castration was completed when a herder bit and ate the testicle. These were known as Rocky Mountain oysters.

California. Just as when the sheep were tagged at the end of the summer, corrals and shearing shoots are built. The sheep are corralled and moved through the shearing shed in an assembly line process, just like the tagging process. Up to 1883, the sheep were sheared twice a year, but the amount of wool was low both times. It increased when one shearing began.[214] Wool is bagged and then trucked either to a warehouse in Texas or to Los Angeles, where it is shipped overseas.

Wool clip ready to send to warehouse.
Photograph courtesy of Elizabeth Espil Mooney.

214 "Report of the Acting Governor of Arizona Territory to the Secretary of the Interior 1881", Arizona State Government Publications, assessed July 21, 2016, azmemory.azlibrary.gov/cdm/ref/collections/statepubs/id/4580.

Trailing Stories

Dolan Ellis related his and his son's experiences trailing the Epsil family's sheep and the Basque sheepherders back in the 60s for three days. They spent time along Ash Creek in the early spring, sharing the sheepherders' campfires. Dolan and his son slept in sleeping bags, "ate fresh mutton that the Basque herders had slaughtered, drank their strong campfire coffee, and had mutton chops and eggs for breakfast that had been cooked over the campfire in heavy black skillets."[215] He related how cold the nights were; each morning, heavy frost could be seen on the grass and ground. The shepherds hung their mutton on trees at night for a couple of reasons: to keep it away from coyotes and for freshness. No refrigeration was needed with the cold temperatures. Coyotes were howling each night, and "they (the coyotes) were actually following and stalking the herd, and we would pick up more and more coyotes each night." The sheepdogs earned their wages as they circled the flock every night, barking to keep the coyotes away. The two sheepherders were Basque and thus spoke very little English, but through sign language, grunts, and Dolan's guitar playing and singing, they were able to communicate. "'Malaguena Salerosa,' a traditional folk song, was their favorite song that I sang...and sang...and sang it again and again."[216]

---

215 Personal communication with Dolan Ellis, July 19, 2016.
216 Ibid.

Others have written about time spent on the trail with various sheep outfits. The earliest the author found is a sixty-five-page handwritten journal kept by Vince Vargas during a ten-day period in May 1967.[217] Sue Peterson[218] wrote of her experiences on the trail with Jean Etchamendy during the fall of 1976 on a hundred-mile stretch of the Heber-Reno Sheep Driveway. She joined him at Young and continued to Tonto Creek as he was bringing his sheep down to the Chandler area of the Salt River Valley. Through four children's books, Cindy Shanks weaves the stories with beautiful photographs of her months following the herders of Dwayne Dobson's sheep.

---

217 Journal stored at the Chandler Historical Museum, Chandler, Arizona.

218 Sue Peterson, "Shepherd of the Open Range," *Arizona Highways* (August 1978), 2–9.

CHAPTER SIX

REASONS FOR THE
SHEEP INDUSTRY DECLINE

There were several reasons for the decline in sheep ranching that can be classified under governmental laws but are divided into immigration laws and other bureaucracy.

Immigration Laws

It is necessary to briefly mention some of the immigration laws that affected the sheep industry to understand the complexity of the problem foreign-born sheep ranchers faced and the restrictions on immigrant workers that helped drive many sheep ranchers out of the business in the late 1990s and early 2000s. The Supreme Court ruled in the 1880s that immigration was a federal issue. Starting in the 1790s, immigration legislation stated that nonindentured white males could become citizens if they lived in the United States two years. Then it was changed to five years then to as high as fourteen years[219].

Ellis Island opened in 1892 and was the primary immi-

---

219   Beth Rowen, "Immigration Legislation", *Infoplease*, assessed October 15, 2016, www.infoplease.com/us/immigration/legislation-timeline.html

gration station for the next sixty-two years. Many of the Basque interviewed stated that this was where their family members entered the United States. The requirement for all immigrants to learn English prior to becoming a citizen was part of the Naturalization Act of 1906. The Expatriation Act of 1907 stated that women must adopt the citizenship of their husbands, and if any woman married a foreigner, they would lose their US citizenship unless their husbands become citizens.[220] In the 1920s, the total number of immigrants was reduced for any country to 3 percent of the nationality based on the 1910 census, then 2 percent of that nationality based on the 1890 census, with a further reduction when that 2 percent figure was based on the 1920 census. There was also a reduction in immigrants from Southern and Eastern Europe.[221]

Immigration laws cut off the flow of young shepherds in 1920 from Basque Country. One article mentioned a Basque from France trying to come in with a group of German engineers. The Germans pretended he was also part of their group of engineers. Once he cleared immigration, he left for the west and presumably to the life of a sheepherder.[222] Highly skilled workers were exempted from new quotas enacted, so the engineers could immi-

---

220 "An Act in Reference to the Expatriation of Citizens and their Protection Abroad [March 2, 1907], *History Central,* assessed October 15, 2016, www.historycentral.com/ documents/Expatriation.html

221 "U.S. Immigration Before 1965", *History,* assessed October 15, 2016, www.history. com/topics/u-s-immigration-before-1965.

222 Michele Strutin, "The Basques: Lords of the Range," *Rocky Mountain Magazine* (July/August 1981), 28–35.

grate, but the herder would have been turned away.[223]

During the 1920s, Congress also passed the Married Women's Act of 1922, which repealed the Expatriation Act of 1907.

The Supreme Court decision in the 1880s, the repeal of the Expatriation Act of 1907, the Married Women's Act of 1922, and Arizona's Alien Act of 1921 should never have been used to take Munds Park from the Echeverrias. That they were used can only be attributed to chicanery by the party wanting the land.

The Immigration Act of 1917, which stated that any immigrant older than sixteen must be able to read a forty-word selection in their own language, seemed strange. It is wondered who at Ellis Island could understand Basque, a language of a very small population group and unrelated to any other language, to ensure they were reading the passage.

Two other pieces of legislation need to be addressed to fully understand the plight of getting workers. The first was because of the agricultural worker shortage during World War II, the bracero program allowed Mexican workers to be hired. While many herders and ranchers' sons were being drafted into the military during World War II, the AWGA would write letters to the Selective Service System requesting a herder or rancher's son have a deferment or be given discharge papers. Qualified workers who understood the everyday

---

223 "Chapter 1: The Nation's Immigration Laws, 1920 to Today", *Pew Research Center*, September 28, 2015, www.pewhispanic.org Retrieved October 15, 2016.

operations of a sheep ranch were few. Many ranchers stated that workers were available but not qualified for the sheep industry. A man whose family had always been in the sheep business knew how to control the sheep using the dogs and could recognize when a sheep was ill and how to treat them. He could do the hard work of building corrals when sheep needed to be moved during the winter grazing period, help with the lambing that during a couple months of each year was a 24/7 operation, and most importantly, was willing to spend many hours alone each summer with the sheep. Without qualified men, many of the sheep ranchers had to reduce their herd sizes. A certain size herd was necessary for making a profit, and some sheep outfits completely sold out.

Ranchers across the West organized to petition the government to have the law changed and allow them to hire Basque sheepherders whose families had been in the business and understood the requirements. California, with large numbers of sheep and sheep ranchers, became the voice for the sheep ranchers. The California Range Association successfully lobbied "to pass the 'sheepherder bills' that allowed ranchers to sponsor herders from the Basque Country."[224]

With a name change, the Western Range Association (WRA) was born in the 1950s. "The WRA was not only active in national politics, but also played an interna-

---

224 "Sheepherders of Northern Nevada A Multimedia Exhibit", assessed October 15, 2016, knowledgecenter.unr.edu/sheepherders/wra.html.

tional role in negotiating with the government of Spain and setting up a sheepherder recruiting office in Bilbao. Back in the United States, the biggest challenge for the WRA was to keep members in compliance with federal regulations governing the importation of herders."[225]

Workers came on a three-year contract. The sheep rancher paid their round-trip airfare. A monthly rate was established for all workers and included room and board. Arrangements would be made with each sheep rancher to ensure his workers got to appointments and any mail delivered. Disputes between the rancher and the worker were resolved through the WRA, with most being about pay. Before the three-year period was up, if the rancher and the worker wanted to continue the work relationship, paperwork was begun for rehire. The worker would return to their country for a minimum of three months before being allowed to return to the United States.

In 1965 major changes occurred in the laws that emphasized visas for reuniting family members and employment categories, but the Labor Department had to ensure that an American worker was not available first to fill the position.

When the economy in Spain improved in the 1970s, it was harder to recruit Basque to come to the United States. The WRA began to look to Mexico, Peru, and even Chile to recruit sheepherders. The Western Range Association

---

225 Ibíd.

helped fill a void in getting qualified workers for the ranchers. It still helps the three remaining sheep outfits today.

Other Governmental Obstacles

The creation of the forest reserves has been discussed, but other governmental obstacles affected the sheep industry as a result of further land deterioration brought on by drought and the perception that overgrazing was still taking place. In 1934, President Roosevelt signed the Taylor Grazing Act into law. The purpose of the act was to safeguard the public lands from overgrazing, provide for the use of the public domain with improvements, and "stabilize the livestock industry dependent upon the public range."[226]

Many "sheep operators who received a grazing permit for public lands increased the stability of their operations and were able to maintain a sustainable operation."[227] Some sheep ranchers solved part of their problem by buying or leasing private pastures. Changes to the Federal Land Policy and Management Act (FLP-MA) of 1976 were the result of changes in social values of the growing population who wished to see wildlife and the environment protected; a balance was needed for all those interested in using the land.

The Western Watersheds Project sued to close the

---

226 "History of Public Land Livestock Grazing", *U.S. Department of the Interior Bureau of Land Management,* assessed October 15, 2016, www.blm.gov/ut/st/en/[rpg/grazing/history_of_public.html.
227 Ibid.

Heber-Reno Morgan Mountain Driveway in 2009 over concerns that one domesticated sheep could expose the bighorn sheep to a respiratory disease that could have dire consequences for the wild sheep. An environmental assessment reached the conclusion that no significant impact would result and authorized the use of four thousand sheep on the Heber-Reno Driveway and eight thousand for the Morgan Mountain Driveway. A driveway must have sheep, even only one, run on it each year to be considered a viably active driveway for future use. The last sheep was trailed on the Heber-Reno in 2011 by the Dobson Sheep Company. The only active driveway is the Black Canyon.

The use of driveways has been a subject of contention between the government and the AWGA clear back into the 1920s and 1930s. One particular conference between those representing the government and members of the AWGA took place in 1939, when it was suggested that the driveways either be eliminated or reduced in width.

According to Louis Espil as told to the *Phoenix Gazette*, August 3, 1979, other regulatory agencies that put up roadblocks were the Department of Labor, Department of Economic Security, Environmental Protection Agency, Department of the Interior, Endangered Species Act of 1973, Department of Agriculture, Division of Wildlife Services, and Occupational, Safety, and Health Administration (OSHA). The Department of Economic Security inspected camps and facilities as they didn't want herders living in sheep trailers or working more than eight hours a day. Human waste was a concern of several of the agencies, as told by one sheep rancher.

With some agencies, endangered species and wildlife take priority, so restrictions on where sheep can graze and water rights were now being questioned. The Mexican gray wolf was reintroduced to the wild in east-central Arizona in the area of grazing for many sheep in 1998. The wolf is a predator of the sheep, and sheep ranchers were not allowed to kill them.

Ranchers said that the Forest Service allowed trees to grow over grasses, thus the grasses disappeared for grazing purposes. Nonnative species were introduced, such as the tamarack or salt cedar. As second homes were built in the forest reserves, more people came to cool off in the summer from the desert heat. Forest fires no longer helped clean the forest of old and dead trees to allow the grasses that in turn allowed livestock to graze.[228]

The government lifted tariffs on importing cheap foreign lamb and wool, and that has taken a toll on the sheep ranchers' profits. Espil and other ranchers compared imported sheep to jerky versus nice, lean meat. Prices are lower for imports, and consumers buy the inferior brand not knowing the difference in taste or quality.

Sheepherders kept moving south to the Casa Grande area due to population pressures in the Phoenix area. Farmers who allowed the sheep to graze their alfalfa fields were

---

228   Oral history interview with Doy Reidhead [with transcript] April 3, 2006 Ecological Oral Histories Collection. Northern Arizona University.

now selling the alfalfa to the dairy men at a higher profit. As farmers sold out to developers who built subdivisions, the dairy cows got pushed out, just like the sheep. Sheep being moved on the roads between pastures in the winter was an inconvenience to drivers, and no one wanted to live near a dairy farm. Many families interviewed said they could not keep up with social attitudes.

Moving sheep across the road before heading to the Verde River.
Photograph courtesy of David W. Schafer Photography.

"People don't realize how important agriculture is in their lives. Without food and fiber, you wouldn't have food on your table and clothes on your back."[229] While

---

229 Personal communication with Irene Aja, various dates, May 2016

wool and sheep are not considered a major part of the Arizona agricultural scene today unless one lives on the Navajo Indian Reservation, there are many benefits gained from raising sheep. The many byproducts are strings for musical instruments, cosmetics (lanolin from the skin), insulin for diabetics, candles, glue, fertilizers, shoes, hat bands, college diplomas, and wool, which is resistant to fire. Military uniforms were made from wool because it is fire-retardant. Sheep help manage the forest by foraging on brush and grass, which would otherwise fuel forest fires. Proper sheep grazing means you don't need to clear the forest or conduct controlled burns that may result in an out-of-control fire. Sheep foraging on undesirable brush and weeds can reduce, if not eliminate, the need for herbicides. In many ways, sheep are more cost effective then raising cotton, which requires more water and land mass to grow the crop.[230]

As we close the history of the sheep industry in Arizona, it is important to remember that individuals and families are involved with sheep in ways other than the traditional way followed for so many years and presented here. One group raises sheep for 4H students so they have a supply of lambs to show in county and state fairs. In Arizona, there are four major sellers for the students, with two of them having between eighty and a hundred sheep. About twelve families have between five and twenty

---

230 "Sheep and Brush Control" and "Sheep and the Environment Series." *American Sheep Industry Association*, assessed October 12, 2016, www.sheepusa.org/ResearchEducation_Literature_SheepAndTheEnvironment.

United States Department of Agriculture Funder/Sponsor. Join a sheep club Twenty sheep to equip and clothe each soldier// HS. [Phila: Breuker & Kessler, Co., or 1918, 1917] *Image. Retrieved from the Library of Congress, https://www.loc.gov/item/00651569/. (Accessed October 22, 2016.)*

sheep for the same purpose. Students must purchase their lambs, and they may cost anywhere from $300 to $1,000. One outfit that sells to the students told the author the sheep is the Hampshire Suffolk, known for its meat, not its wool. Before the student buys the lambs, the animal is sheared, vaccinated, and checked for any deformities. The outfit looks for sheep with good carcasses as that is what the animal will be judged on at the fair.

Those who train dogs for herding are the second group keeping a few sheep. The Arizona Herding Association (AHA) has the "objective to promote herding among all breeds (dogs)."[231] Those who train the dogs keep a small flock of sheep for the dog owner to work with as the dog is trained. The AHA sponsors American Kennel Club-licensed herding trials. These trials are in small enclosures for dogs just learning and can include activities for the more experienced dog out in an open field. The author attended one of these events in May 2016 and witnessed the dogs in action and talked to many of the dog owners. While some dogs were there more for a bonding time between the owner and dog, other dogs would actually be sold to herders.

Many families gave the author permission to tell their stories, but not all families were able or willing to participate. It is felt that a good portion of the history of the sheep industry has been told through personal interviews and what has been collected from historical

---

231 "About Us," *Arizona Herding Association,* assessed October 12, 2016, arizona-herding.com.

Arizona's 'Make It With Wool' Contest Winner, Lois Knudson, 2015 (above), and State Director, Rali Burleson (below), who told the author she also is "promoting the beauty and versatility of wool". *Photographs courtesy of Barbara Trainor Photography.*

documents.

Sheep were important to the early economy of Arizona. In 1906, there were 150 sheep outfits belonging to the AWGA. In 1961, a list of the amount of wool produced from 1944 to 1961 showed sixty-three outfits. While Louis Espil predicted that sheep would be totally gone from the landscape by the 1980s, two families and three sheep outfits are still holding on in 2016, continuing the trailing of sheep on at least part of the Black Canyon Driveway. The days of trailing and raising large bands of sheep will be gone as the last two families leave their three businesses; sheep will still be found in small numbers on family farms and on the Navajo and Hopi Indian reservations.

Dog training with sheep.
Photograph courtesy of author.

Bibliography

A Historical and Biographical Record of the Territory of Arizona.
   Chicago: McFarland and Pool, 1896.

"About Us", *Arizona Herding Association,* assessed
   October 12, 2016, arizonaherding.com.

"An Act in Reference to the Expatriation of Citizens and their Protection
   Abroad [March 2, 1907], *History Central,* assessed October 15, 2016,
   www.historycentral.com/documents/Expatriation.html

Arizona Wool Growers Association, NAU.MS.233, Northern Arizona
   University, Cline Library Special Collections and Archives Dept.

Balza, Joan. "Last of the Old Sheep-Drive Routes: The Heber-Reno Trail."
   Arizona Highways. (November 1986.)

Benanav, Michael. "The Sheep are like our Parents."
   The New York Times Travel Section. July 27, 2012.

Bolton, Herbert Eugene. *Rim of Christendom: A Biographer
   of Eusebio Francisco Kino Pacific Coast Pioneer.*
   Tucson: University of Arizona Press, 1986.

------- *The Padre on Horseback A Sketch of Eusebio Francisco Kino,
   S. J. Apostle to the Pima.* Chicago, IL: Loyola Press, 1982.

Breeds of Livestock, Department of Animal Science.
   "Breeds of Livestock-Navajo-Churro Sheep."
   Oklahoma State University. Updated October 22, 1996.
   www.ansi.okstate.edu/breeds/sheep/navajochurro.

Brooks, Juanita. "Jacob Hamblen." *Arizona Highways* 19, no. 4 (1943).

Carlson, Frances. "James D. Houck: The Sheep King of Cave Creek."
   The Journal of Arizona History 21, no. 1 (1980).

"Chapter 1: The Nation's Immigration Laws, 1920 to Today",
   Pew Research Center, September 28, 2015,
   www.pewhispanic.org Retrieved October 15, 2016.

Cline, Platt. *They came to the Mountain.*
   Flagstaff: Northern Arizona University, 1976.

Coues, Elliott. *On the Trail of a Spanish Pioneer: The Diary and Itinerary
   of Francisco Garces.* New York, NY: F.P. Harper, 1980. II

Cremony, John C. *Life among the Apaches.* Glorieta: New Mexico:
   The Rio Grande Press, 1969.

Croxen, Fred. "History of Grazing on Tonto." Presented at the Tonto
   Grazing Conference in Phoenix, Arizona November 4–5, 1926.

Davis, W. W. H. *El Gringo*, or New Mexico and her people.
   New York: Harper & Bros, 1857.

French, Doris. comp. and ed. "Flora Locarnini Erramuzpe,"
   Arizona National Ranch Histories of Living Pioneer Stockman, 2003.

Gerald A. Doyle & Associates, P.C. "Verde River
   Sheep Bridge," Phoenix, AZ: July 1987.

Douglas, William and Jon Bilbao. *Amerikanuak: Basque in the New World.*
   Reno, Nevada: University of Nevada Press, 1976.

Embach, Harry B. "Jose Antonio Manterola—A Real Sheepman."
   Arizona Stockman (May 1950).

Euler, Robert Clark. "A Half Century of Economic Development in
   Northern Arizona 1863-1912." (master's thesis, Arizona State
   College, 1947).

Farish, Thomas. Farish Arizona History. Phoenix, AZ:
    Manufacturing Stationers, 1920.

Fuchs, James R. *A History of Williams, Arizona 1876–1951*.
    Tucson, AZ: University of Arizona Press, 1955.

Garate, Donald. "Arizona, A Twentieth-Century Myth."
    Journal of Arizona History 46, no.2 (2005).

Gregg, Josiah. *Commerce of the Prairies*,
    Philadelphia, PA: Lippincott, 1962.

Hamilton, Patrick. *The Resources of Arizona*. Territory: Prescott, 1881.

Hammond, Geogre Peter. *Don Juan de Onate and the Founding of
    New Mexico*. Santa Fe, New Mexico: El Palacio Press, 1927.

Haskett, Bert. "History of the Sheep Industry in Arizona."
    Arizona Historical Review 7 (1936).

Hinton, Richard. *The Hand-Book to Arizona: Its Resources, History,
    Towns, Mines, Ruins and Scenery*. San Francisco: Payot,
    Upham & Co., New York: American News Co., 1878.

"The History of the Navajo Sheep Project Organization",
    Navajo Sheep Project, assessed October 1, 2016,
    www.navajosheepproject.com / nsphistory.html.

"History of Public Land Livestock Grazing", *U.S. Department of the
    Interior Bureau of Land Management*, assessed October 15, 2016,
    www.blm.gov / ut / st / en / [rpg / grazing / history_of_public.html.

Hodge, Hiram. 1877. *Arizona As It Is*. Compiled from notes of Travel
    during the years 1874, 1875 and 1876. New York, NY: Hurd and
    Houghton, Boston, MA: H.O. Houghton and Company, 1877.

Jackson, Orick. *The White Conquest of Arizona*. Los Angeles:
    West Coast Magazine, Grafton Co., 1908, pg. 45

James, George Wharton. *Indian Blankets and Their Makers.* Glorieta, NM: Rio Grande Press, 1914.

Kessell, John. Friars, *Soldiers and Reformers: Hispanic Arizona and the Sonora Mission Frontier, 1767–1856.* Tucson: University of Arizona Press, 2016.

Kildare, Maurice. "Jose Chavez The Man Who Refused to Die." *Real West* (March 1968).

Kyser, Dawn. "End of an Era: Red Tape Strangling State Sheepherders." *Phoenix Gazette.* August 3, 1979.

Mallea-Olaetxe, Joxe. "Carving Out History: The Basque Aspens." *Forest History Today* (Spring/Fall 2001).

"Navajo-Churro Sheep", The Livestock Conservancy. Heritage Breeds. accessed October 1, 2016, www.thelivestockconservancy.org

"The Navajo, Sheep, and the Federal Government", *Native American Netroots,* November 10, 2011, assessed October 1, 2016, www.nativeamericannetroots.net/diary/1136

Nourse, Ruth. "First Lady of Sheep-Growing Clan Thinks and Cooks in Two Languages." *The Arizona Farmer-Ranchman* (October 19, 1968).

Oral History Interview with Senator Henry Fountain Ashurst [includes transcript] May 19, 1959 and October 9, 1961, Northern Arizona University, Cline Library.

Oral History Interview with Mamie Fleming [includes transcript], March 7, 1953, Northern Arizona University, Cline Library.

Oral History Interview with Joe Manterola, June 18, 2008. Colorado Plateau Archives, Northern Arizona University, Cline Library.

Oral history interview with Doy Reidhead [with transcript] April 3, 2006 Ecological Oral Histories Collection. Northern Arizona University, Cline Library.

Peplow, Edward H. Jr. *History of Arizona, Volume II.*
New York: Lewis Historical Publishing Company,1958.

Peterson, Sue. "Shepherd of the Open Range."
Arizona Highway 54 (August 1978).

Pollock, Paul W. "Fermin Echeverria." *American Biographical*
Encyclopedia, Volume 1. Phoenix, AZ: Paul W. Pollock, 1967.

-----"Sanford W. Jaques." *American Biographical Encyclopedia,*
Volume 1. Phoenix, AZ: Paul W. Pollock, 1967.

-----"Michel Joseph O'Haco." *American Biographical Encyclopedia,*
Volume 1. Phoenix, AZ: Paul W. Pollock, 1967.

-----"M. P. (Pete) Espil." *American Biographical Encyclopedia,*
Volume 2. Phoenix, AZ: Paul W. Pollock, 1969.

----- "Louis Albert Espil." *American Biographical Encyclopedia,*
Volume 3. Phoenix, AZ: Paul W. Pollock, 1974.

----- "Jean Etchamendy." *American Biographical Encyclopedia,*
Volume 5. Phoenix, AZ: Paul W. Pollock, 1981.

Portrait and Biographical Record of Arizona.
Chicago: Chapman Publishing Co, 1901.

"Preservation An Endangered Breed", *Navajo-Churro Sheep Association,*
assessed October 1, 2016, Navajo-churrosheep.com/index.html

Reeve, Frank D., ed. "The Sheep Industry in Arizona, 1903."
New Mexico Historical Review 37, no. 3. (July 1963).

Reeve, Frank D., ed. "The Sheep Industry in Arizona, 1905–1906."
New Mexico Historical Review 37, no. 4 (October 1963).

Reeve, Frank D., ed. "The Sheep Industry in Arizona, 1905–1906."
New Mexico Historical Review 38, no. 1 (January 1964).

Reid, Betty. "Arizona's American Indians – Sheepherder", assessed October 1, 2016, www.archive.azcentral.com/news/native-americans/?content=sheepherder.

"Report of the Acting Governor of Arizona Territory to the Secretary of the Interior 1881", Arizona State Government Publications, assessed July 21, 2016, azmemory.azlibrary.gov/cdm/ref/collections/statepubs/id/4580.

Rowen, Beth. "Immigration Legislation", *Infoplease,* assessed October 15, 2016, www.infoplease.com/us/immigration/legislation-timeline.html

"Sheep and Brush Control" and "Sheep and the Environment Series." *American Sheep Industry Association,* assessed October 12, 2016, www.sheepusa.org/ResearchEducation_Literature_SheepAndTheEnvironment.

"Sheepherders of Northern Nevada A Multimedia Exhibit", assessed October 15, 2016, knowledgecenter.unr.edu/sheepherders/wra.html.

Sheffer, H. Henry III and Alger, Sharyn R. *The Pleasant Valley War Cattle and Sheep Don't Mix.* Apache Junction, AZ: Norseman Publications, 1994.

Short, Cathy. "Lamb to Loom," *First American Art* no. 9, Winter 2015.

State of Arizona. "Arizona's Alien Land Act." *Revised Code of Arizona,* 1928.

Strutin, Michele. "The Basques: Lords of the Range." *Rocky Mountain Magazine* (July/August 1981).

Towne, Charles Wayland and Edward Norris Wentworth. *Shepherd's Empire.* Norman: University of Oklahoma Press, 1945.

Turner, Jim. "How Arizona did NOT Get its Name."
  Arizona Historical Society. Archived from the original
  on August 1, 2007. Retrieved May 15, 2016.

U.S. Department of Agriculture, Bureau of Animal Industry.
  *Special Report on the History and Present Condition of the Sheep
  Industry of the United States*, prepared under direction of Dr.
  D. E. Salmon, Chief, Bureau of Animal Industry, Washington,
  D.C.: Government Printing Office, 1892, Part II.

"U.S. Immigration Before 1965", *History*, assessed October 15, 2016,
  www.history.com/topics/u-s-immigration-before-1965.

Vinson and Dwayne Dobson Interview, *Chandler*, September 12, 2006
  www.chandleraz.gov/content/Oral_history_collection_list.pdf

Wentworth, Edward Norris. *America's Sheep Trails*.
  Ames, Iowa: IA State College Press, 1948.

Wyllys, Rufus K. Arizona: *The History of a Frontier State*.
  Phoenix, AZ: Hobson & Herr, 1951.

INDEX

Index

Range Association
War Production Board, 208
Washington, George
  (U.S. president), 1
Western Range Association
  (WRA), 105, 238-239
Wheeler, Ike, 53
Whipple, Amiel, Lieutenant, 16
White, 15
White House, 1
White Mountain Electric
  Cooperative, 81
White, Josiah, 15
Whittemore, Beulah A., see
  Jaques, Beulah A. Whittemore
Wickenburg, AZ, 89, 103-104,
  106, 112, 116, 120-121, 126-
  127, 137, 140, 142, 154
Wilbur, George, 32
Williams, Mr., 26
Williams, AZ, 39, 41, 69, 89-
  90, 93, 104, 134, 142-143,

147, 149, 157, 176, 183
Wine Glass Ranch, 135
Winslow, AZ, 52, 62, 74,
  110, 120-122, 126, 128, 134,
  137, 154-155, 168, 176
Wolfswinkel, Kathy, see
  Aleman, Kathy Wolfswinkel
Woodbridge, Jerry, 50
Woods, 73
Woolf, Ed, 164

Yaquis, 15
Yaeger, Henry, 76-77
Yavapai County, 28, 40-41
Yeager, Bill, 53
Yrissari, Manuel, 36
Yuma, AZ, 13-15, 27, 30, 161

Zunis, 172